How To Write a Lab Report

Jerome N. Borowick
California State Polytechnic University, Pomona

Prentice Hall
Upper Saddle River, New Jersey Columbus, Ohio

Cover art: Katherine Hanley
Editor: Stephen Helba
Production Editor: Alexandrina Benedicto Wolf
Cover Design Coordinator: Karrie Converse-Jones
Cover Designer: Jason Moore
Production Manager: Matthew Ottenweller
Marketing Manager: Chris Bracken

This book was set in Century Schoolbook by Carlisle Communications, Ltd., and was printed and bound by The Banta Company. The cover was printed by Phoenix Color Corp.

Pearson Education
Upper Saddle River, New Jersey 07458

Printed in the United States of America

10 9 8 7 6 5 4 3 2 1

ISBN: 0-13-013562-3

Prentice-Hall International (UK) Limited, *London*
Prentice-Hall of Australia Pty. Limited, *Sydney*
Prentice-Hall of Canada, Inc., *Toronto*
Prentice-Hall Hispanoamericana, S. A., *Mexico*
Prentice-Hall of India Private Limited, *New Delhi*
Prentice-Hall of Japan, Inc., *Tokyo*
Prentice-Hall (Singapore) Pte. Ltd., *Singapore*
Editora Prentice-Hall do Brasil, Ltda., *Rio de Janeiro*

Table of Contents

SECTION I: PRINCIPLES OF TECHNICAL WRITING

SECTION II: ELEMENTS OF THE LAB REPORT

Preface

As an undergraduate engineering student, I completed many laboratory classes that required me to regularly submit laboratory reports. Although my instructors discussed what should be in these reports, they rarely gave any instructions on how they should be prepared.

This situation undoubtedly created frustration due to my inability to understand the purpose of these reports and how to achieve this purpose. Consequently, it became a time-consuming task where the rewards rarely seemed to justify the time spent writing these laboratory reports.

The importance of the numerical results obtained from the laboratory experiments was obvious to me: they demonstrated the principles of engineering and science that I had learned in the classroom. The importance of learning to use laboratory equipment was also obvious to me. However, the importance of learning to write effective laboratory reports was not as obvious. I did not realize at that time that writing these reports would lay the foundation for the technical reports I would later be writing as a professional.

For most engineering and science students, writing laboratory reports is their introduction to writing technical reports. For this reason, I have written this text to help prevent some of the frustrations that I experienced as a student. It demonstrates the methods for writing effective laboratory reports, and provides a basis for writing effective technical reports as a professional.

Figure P-1 shows the "Instructor Requirements for Laboratory Reports" matrix. This should be completed under the direction of your instructor before you write your report.

I thank the hundreds of students and the engineering faculty at Cal Poly, Pomona who have, since 1981, encouraged and helped me to write this text. I also acknowledge the many students who have graciously permitted me to use their class assignments in this text.

Jerome N. Borowick

INSTRUCTOR REQUIREMENTS FOR LABORATORY REPORTS
For each class indicate the sequence of sections to be included
(See sample ME 299)

Item **Dept. & Class No.**	ME 299									
Cover Materials (Chap. 6)										
Standard Lab. Folder (p. 53)	X									
Title Sheet (p. 53)										
Table of Contents (p. 55)	X									
Binding	staple									
Other										
Begin. of the Report (Chap. 7)										
Summary (p. 57)	11									
Object., Purp., or Intro. (p. 58)	1									
Introduction (p. 58)										
Nomenclature/Def. (p. 59)	2									
Symbols (p. 59)										
Other										
Body of the Report (Chap. 8)										
Theory (p. 61)	3									
Assumptions (p. 62)	7									
Sample Calculations (p. 63)	4									
Tested Items (p. 64)										
Equip. and Apparatus (p. 65)	5									
Process Description (p. 65)	6									
Laboratory Procedure (p. 65)										
Data and Calc. (p. 66)	8									
Other										
Ending of the Report (Chap. 9)										
Results and Conclu. (p. 69)	9									
Discuss. of Results (p. 70)	10									
Bibliography (p. 71)										
Appendix (p. 72)	12									
Other										

FIGURE P-1

SECTION I

Principles of Technical Writing

CHAPTER 1

An Introduction to Writing Lab Reports and Technical Writing

THE DEFINITION OF TECHNICAL WRITING

Technical writing is the written communication of engineering and scientific ideas, concepts, and data presented objectively, logically, and accurately. The writing is clear and concise. Graphics are usually included to assist the reader in understanding the subject matter. The reader should be convinced of the conclusions based on the information presented. The recipient will usually use the report or document to perform a task, make a decision, solve a problem, or acquire information and knowledge.

THE IMPORTANCE OF TECHNICAL WRITING

As a technical professional, you will have the opportunity to perform many functions: design, analyze, research, manufacture or construct, test, and manage. In the performance of these functions, many technologists, engineers, and scientists new to the work environment discover that they may spend as much as 50 percent of their time writing reports and documents that discuss the results of their work.

Your technical competence will be important to the successful completion of the projects you work on. In today's high-tech industrial society, it is likely that your work on these projects will be communicated and coordinated with other professionals, government agencies, clients, and managers who depend on your results. Therefore, the successful completion and profits of any project may be jeopardized without effective communication between professionals that discusses progress and problems.

A few years ago, the Chief Executive Officer of Lockheed-California said the following to a group of engineering educators:

> From a purely technical and scientific standpoint, your engineering colleges are sending us the best-educated, best-prepared new engineering graduates we have had. We are strained to challenge their technical and quantitative abilities. However, these same graduates have one glaring deficiency: they can't write. I want you to send me engineers who can WRITE! WRITE! WRITE! WRITE! WRITE!

Messages similar to this one are repeated over and over by leaders of industry, government, and academia. The importance of your ability to technically write well cannot be understated.

From a personal standpoint, your peers and colleagues sometimes will judge your technical competence by evaluating the effectiveness of your writing. Frequently,

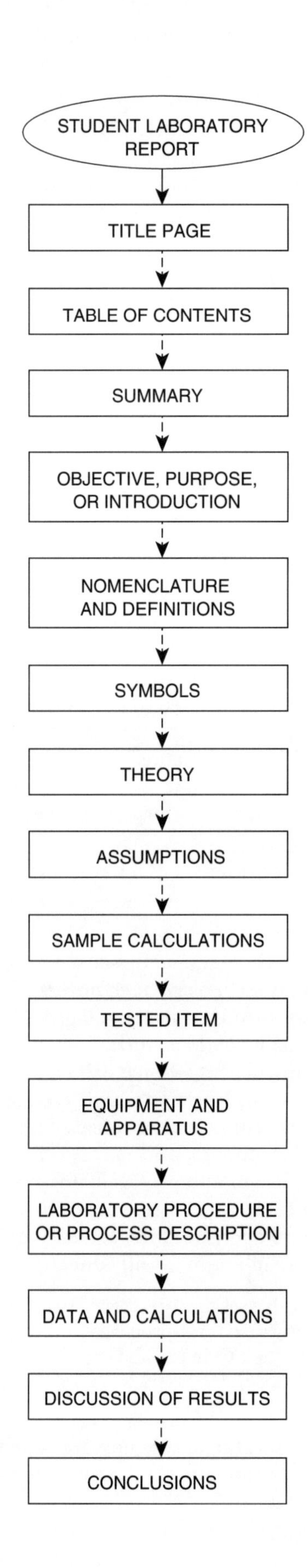

FIGURE 1–1
Elements of a Student Laboratory Report

performance reviews by supervisors are subtle reflections of your ability to write. Your reports and documents may be your only form of communication with clients, government agencies, and professionals at other facilities and companies. Your reputation as a professional depends not only on your ability to perform well, but also on your ability to write well.

THE STYLE OF TECHNICAL WRITING

Effective technical writing is objective, clear, concise, and convincing. This is accomplished when the style is descriptive and quantitative; that is, when the writing includes details and uses facts, data, measurements, and statistics. All pertinent information is presented in a manner that can be objectively evaluated and concluded by the reader. Imprecise, judgmental, emotional, and editorial words such as *many, undue,* or *annoying* are avoided to deter readers from interpreting the magnitude of the meanings of these words based on their own personal experiences. Only information that is relevant to the purposes of the reports is included. Information that is interesting, but not pertinent, is excluded. The purpose of the communication determines the tone (manner and attitude of expression).

College writing addresses an audience that knows more than the writer and that needs a basis for determining the level of understanding of the writer (student). This writing discusses already proven concepts and ideas supported by selected facts and data. By contrast, technical writing addresses an audience that knows less than the writer and that needs information. This writing discusses unknown concepts and ideas supported by all pertinent facts and data. Therefore, the information presented in technical writing must be persuasive by including facts, data, and analysis rather than by convincing argument.

STUDENT LABORATORY REPORTS

In a scientific laboratory, the validity of a hypothesis can be tested from a given set of facts or data. This is known as the scientific method; beginning with the known, the unknown can be discovered.

Because the scientific method plays a significant role in educating engineering and science students, this chapter discusses the reports that result from students' experiences in the laboratory.

CONTENTS OF STUDENT LABORATORY REPORTS

Student laboratory report requirements vary from instructor to instructor, but a formal laboratory report may contain the following elements shown in the order in which they typically appear (see Figure 1–1). Figure 10–1 presents a complete student laboratory report.

Title Page
Table of Contents
Summary
Objective, Purpose, or Introduction
Nomenclature and Definitions
Symbols
Theory

Assumptions
Sample Calculations
Tested Item
Equipment and Apparatus
Laboratory Procedure or Process Description
Data and Calculations
Discussion of Results
Conclusions

LISTS USED IN THIS TEXT

Lists, when used to present information, are effective for organizing your material into a series of discrete items for discussion and for helping readers to comprehend and remember the information presented.

- When the items in a list are preceded by numbers or letters in sequence (e.g., 1, 2, etc., or a, b, etc.), the sequence of operation is important and should be followed in the order presented. This convention is typical for procedures and other chronological series of events.
- When the items in a list are preceded by bullets, such as in this list, the items are unrelated and can be addressed in any sequence. However, the items are usually included in order of descending importance.

DEFINITIONS OF TERMS USED IN THIS TEXT

For clarity, the following definitions of terms used in this text are provided:

- **Professional** includes engineer, scientist, architect, and other technical professionals.
- **Reader** includes any member of your intended audience.
- **Heading** includes the functional heading of a section that discusses the function of the subject matter (e.g., theory, discussion of results).
- **Section** includes the paragraphs where functional discussion of the subject matter (e.g., theory, discussion of results) is found.

GROUP REPORTS

Frequently, reports are written as a concerted effort of a team of several students. This experience can be very rewarding and challenging because it is an opportunity to learn from, and contribute to, the expertise of the other students in the group.

A group leader, either appointed by the instructor or chosen by the other students, can define the scope of responsibility for each student in the group to avoid duplication of effort.

The group leader needs to organize and coordinate the writing of the report and to direct the efforts of the students toward preparing a consistent, well-written report. Each student is responsible for cooperating with the other students and learning from the leadership and judgment of the group leader.

Unfortunately, special problems such as irreconcilable viewpoints between students and shirking of responsibilities sometimes arise when writing group reports. These problems need to be resolved tactfully by the group leader in a manner that is workable and deemed acceptable by all group members. Occasionally, the instructor resolves these problems and may penalize a group member. When this occurs, it should not be perceived as a win or loss by individual group members, but rather as an action intended to benefit the group and, therefore, all members. In spite of its potential drawbacks, writing group reports is typically used by instructors and should be an experience to be eagerly anticipated.

WORD PROCESSORS AND WORD PROCESSING PROGRAMS

The writing process has been revolutionized by word processors and word processing programs. Writers can easily and quickly revise, edit, and check spelling with a spell-checking program. Headings can be enlarged and differentiated from the body of the text with different fonts and styles. Graphics can be inserted into the body of the text. The time required to write a report is reduced considerably. First drafts can use the same structure and format as the intended final drafts. As a result, writing has become more gratifying for many professionals, and the quality of their work has improved. You are encouraged to learn to use a word processor or word processing program as soon as possible.

Computer programs are available to check grammar and readability of writing. These programs can help you improve your writing, but they are merely tools; the responsibility for learning to write is yours. Spell-checking programs are excellent for eliminating spelling errors, but they cannot differentiate between correctly spelled homonyms that have unrelated meanings for the intended words. For example, "The break must be repaired" is concerned with a structural deficiency, whereas "The brake must be repaired" is concerned with a mechanical deficiency. Because these homonyms can mislead the reader and result in an unintended message, the writer still needs to proofread carefully.

RESPONSIBILITY AND ETHICS

As the creator of original writings, you will have the opportunity to inform your readers of your findings. These readers will rely on your writings to make decisions and determine courses of action. This imposes a serious responsibility to state your findings as objectively and honestly as possible without bias to your own purposes. As a professional, you have an ethical responsibility to report your findings completely and accurately regardless of the consequences. Keep in mind that results reported in an unbiased, complete, and accurate report will enhance your reputation and encourage other professionals to rely on your writings, opinions, and judgments.

CHAPTER 2

The Lab Report Writing Process

Effective writing is time-consuming. Students commonly err by allowing inadequate time to prepare their reports properly. The technical laboratory report writing process should not be a concentrated effort; rather, it should be several smaller efforts separated in time to help you organize your ideas. It is most efficient to begin writing the components of a report as you complete the laboratory phase of your experiment so that ideas are fresh in your mind.

Waiting to write your report the night before it is due does not give you enough time to revise and edit your report.

THE SIX-STAGE REPORT WRITING PROCESS

The six-stage process described next (see Figure 2–1) is recommended for writing your report:

1. Define the purpose of the experiment. What did it intend to accomplish?

This information is usually presented in your laboratory manual near the beginning of the discussion of the experiment.

2. Understand the expectations of your instructor. What sections of the report does your instructor emphasize?

- Does your instructor stress the importance of understanding the laboratory procedure and equipment so that correct data will be collected in the laboratory? Are clearly presented, accurate calculations more important than the data collected? Or, are discussion and critical thinking sections the most important?
- How much detail does your instructor expect you to include? Some instructors prefer collected data, calculations, and conclusions only, while others require a formal report. Some instructors require a formal report from only one member of the group that performed the experiment.
- Does your instructor have a specific format to follow?
- As you write your report, remember that the instructor knows significantly more concerning the experiment than you do, and that your objective is to demonstrate to your instructor that you understand the purpose of the experiment and what it accomplished. This is different than technical writing in general, where the reader knows significantly less than you do and your objective is to help the reader understand your purpose and what was accomplished.

FIGURE 2–1
The Technical Writing Process

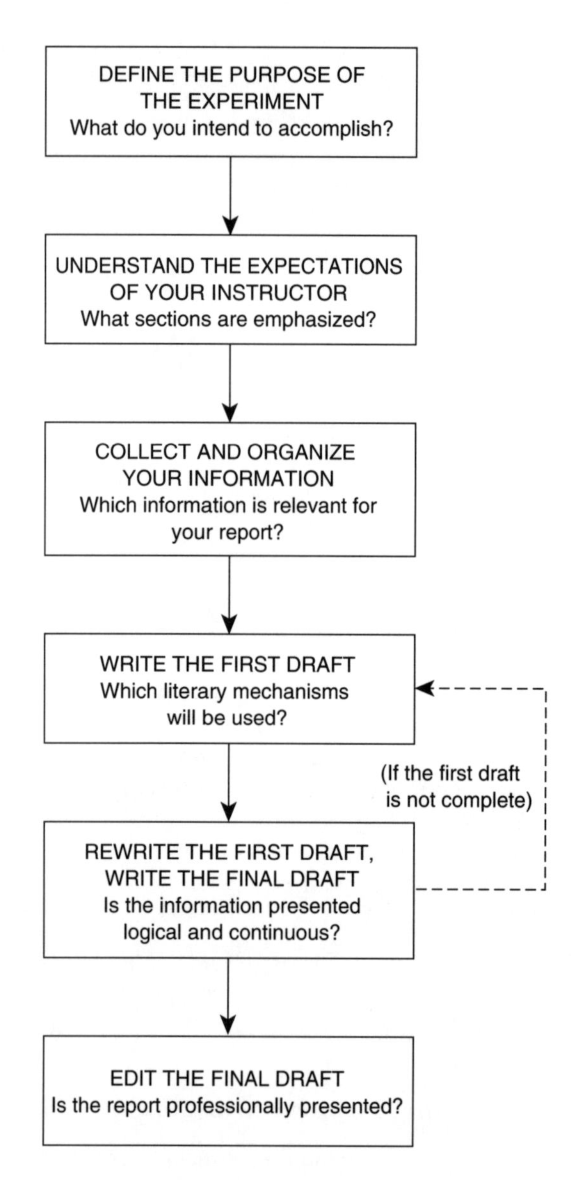

3. Collect and organize your information. You will spend your time in the laboratory gathering data and discussing the procedure and results with the other members of your group.

- **a.** Gather the information in your laboratory manual concerning this experiment, the data you have collected, and the notes and calculations generated while performing the experiment.
- **b.** Gather needed material, such as the theory, available from other sources.
- **c.** Determine which information should be included in your report to meet the expectations of your instructor. Place non-relevant information aside until after your report is graded and returned to you by your instructor (seemingly non-relevant information may help validate your instructor's comments).
- **d.** Separate information into the following four categories: given and available information, data collected in the laboratory, calculations, and notes concerning the performance and outcome of the experiment.

  Determine which sections are required by your instructor to be included in your report. If your instructor has not recommended or identified required sections, use the suggested sections discussed in Chapter 1 and Figure 1–1. Separate the information within the above four categories into these sections.

e. Determine the sequence required by your instructor for including these sections in your report. If your instructor has no recommended or required sequence, use the suggested sequence discussed in Figure 1–1. Arrange these sections in this sequence. This becomes the preliminary draft of your report.

4. Write the first draft. Several literary mechanisms can be used effectively to impart the different aspects of information presented in the report. Each section may use one, or a combination, of the following:

- Descriptions create visual images, and explanations interpret occurrences and phenomena. Frequently an explanation will accompany a description. Objects, physical concepts, and outcomes are ordinarily described and explained. The equipment and apparatus, and discussions and results sections of your report can be descriptions and explanations.
- Chronologies present a series of events. Procedures and process descriptions are ordinarily presented chronologically. Many descriptions and explanations are also presented chronologically. Summaries are usually chronologies.
- Data and calculations and results sections evaluate the laboratory data. These sections of the report determine the conclusions.

Begin by writing the easiest section first. Many students begin with the graphics (i.e., data and calculations, graphs) before writing the text of the report. Technical descriptions and procedures are also good starting places since these do not require a knowledge of the outcome of the experiment (in fact, these are sometimes written prior to performing the experiment). The conclusion and summary are usually the last sections to be written since these require you to understand the results of the experiment and the contents of the report.

You may begin Stage 5 before the first draft of all sections is completed if you are having difficulty writing on one or more of the sections (see "Coping with Writer's Block" below). Writer's block is a common problem even for experienced writers. However, remember that you must eventually return to Stage 4 and complete those sections.

Before you proceed to the next stage you should go on to other unrelated tasks for several days. This will give you a fresh perspective and enable you to review and rewrite your draft in a more objective manner.

5. Rewrite the first draft. Review the logic and continuity of your first draft.

- Students writing first drafts frequently accept the reality that the instructor has more knowledge than they do. Therefore, has adequate information been included to demonstrate to your instructor that *you* understand the purpose of the experiment and what is accomplished?

  Sometimes you may realize that you have omitted a key concept and you may need to add an entire paragraph; other times you may need to add only a phrase or a modifier.
- Is information repeated unnecessarily? Is information included that is irrelevant? Review the ease of readability, accomplishment of purpose, and accuracy.
- Is your writing concise and to the point? Are your style and tone appropriate?
- Have acronyms been defined? Will the instructor understand your abbreviations?
- Are the grammar, spelling, and calculations correct?
- Have you avoided cliches and slang?

Revise your draft until you are confident that the instructor can understand your report without further explanation. Most reports are revised several times before writing the final draft.

6. Edit the final draft. The presentation of your report should be professional when it is submitted to your instructor.

Structure and format:

- Are the entries on the standard laboratory folder or title page completely and neatly entered?
- When required, does the report follow the standard format requested by your instructor?
- Are the margins, headers, footers, and headings esthetically balanced with each other and the text material?

Readability and accuracy:

- Are there any grammatical, spelling, numerical, or typographical errors?
- Are the pages numbered and placed in the proper sequence?
- Do the titles of the graphics represent their content? Are the graphics referenced in the text?
- Is the print dark and large enough to read easily?

COPING WITH WRITER'S BLOCK

Students frequently have difficulty organizing their thoughts and, therefore, beginning writing. This occurrence, commonly known as writer's block, is temporary and should not be attributed to an inability to write. Several different techniques are effective for students working through writer's block. The most commonly used follow:

- Using a thesaurus, work through a complex concept by finding the appropriate words that express your ideas. Then, organize these words into phrases, these phrases into sentences, and these sentences into paragraphs. Revise the passage until it clearly expresses your thoughts.
- Review the parts of the report that are written to determine the unfinished subject matter that needs to be addressed.
- Write other sections of the report until their completion clarifies the purpose of the unfinished section. Then, return to the unfinished section.
- Obtain and read other relevant material to give you new ideas.
- If available, request the help of another group member or a student who is familiar with the laboratory experiment for a fresh perspective to help you break writer's block. Avoid making judgments and carefully consider any suggestions.
- Brainstorm with other students familiar with your experiment. Consider any approaches that have been successful in the past.

ADVICE FOR NON-NATIVE WRITERS

Students whose first language is other than English frequently have anxieties about technical writing. If you are one of these students (or even if you are an experienced writer), consider that all experimental report writing requires the following procedures:

Define the purpose of the experiment.

Collect information and data.

Organize the report.

Analyze the information and data.

Develop graphics.

Write the text of the report.

Proofread.

Establish the format and visual design.

Non-native writers are usually as capable as native writers in all these procedures except, because of their bilingual background, writing and proofreading the text of the report. This difference in background can be overcome by following this procedure:

1. Determine the purpose of the experiment and the expectations of your instructor.
2. Carefully select the useful information and data for inclusion in your report. Do not include any nonrelevant information and data in your report, because it diverts the instructor from your intended objective.
3. Organize the structure of your report and select relevant headings.
4. Fully develop your ideas in the text material, but express them concisely to reduce the number of writing errors.

   Use a language dictionary to help you translate your thoughts into English. Use a thesaurus to help you select the appropriate words to express your ideas. When you are uncertain of the proper use of a word, check its literal meaning in an English dictionary. Ask an experienced writer to help you select the appropriate nouns for technical configurations and concepts (e.g., slotted hole, tongue-and-groove, over-the-center).
5. Support the written text material with calculations and graphics whenever possible. Carefully introduce and label all nontext material.
6. When you have completed your report to the best of your ability, have it proofread by an experienced writer. Carefully study the revisions, and review them with the proofreader.
7. The writing difficulties of non-native writers are typically caused by the improper use of words and idioms and the inability to apply the rules of English grammar appropriately, even though these rules may be understood. Several days after the final draft is completed, you can use this understanding to your advantage by proofreading your own text material to study your application of these rules.

Most schools have a learning resource center or other tutorial service for students who need special assistance with their writing. The cost for services is usually minimal and sometimes may be free. As a student, take advantage of the opportunity to use these services whenever your classes require writing.

Writing improves with practice for all students. Give yourself the opportunity to write whenever you can. Write letters to friends and relatives, keep a diary, and take writing classes. Have your writing assessed by a writing professional whenever possible. With practice, you will gain confidence, and your ability to write effective technical reports will increase.

KEY CONCEPTS

- Effective report writing is time-consuming. Sufficient time should be scheduled to write your report. Preparing the first draft of your report as the experiment progresses expedites the writing.
- An effective writer understands the purpose of the report and the requirements of the instructor.
- It is important to organize your material before you begin writing.

- Selecting the most effective literary mechanism helps the reader comprehend your information.
- Writer's block occurs for all writers. It is overcome easily when you realize that writer's block is temporary. Do not attribute its occurrence to your inability to write.
- An experienced writer should proofread your final draft for clarity and ease of comprehension.

CHAPTER 3

Principles of Clear Lab Report Writing

Technical writing should effectively communicate an idea, a concept, or information. This text teaches you the fundamentals of communication with a purpose of writing clearly and concisely. The principles of grammar and sentence structure are discussed in this text only to the extent that they have an impact on effective technical writing.

This chapter discusses the principles of using a natural style to make your report more understandable, communicating effectively by being direct, writing convincingly by using proper expression, writing effective instructions, and using the proper verb tense for your reports.

Because the principles of laboratory report writing extend to all types of technical writing, the principles and examples shown in this chapter and text relate to all types of technical writing.

KEEPING A NATURAL WRITING STYLE

A natural writing style will increase the readability of your reports and eliminate the need for repeat readings to understand your message. Follow the guidelines in this section (see Figure 3–1) to increase the readability of your writing:

Structure

1. Use lists with headings and introductory remarks. Identify the purpose of the list using headings and introductory remarks. Numbers or letters in sequence (e.g., 1, 2, etc., or a, b, etc.) precede the items in the list when the sequence of operation or chronology of events is important. Numbers or letters also provide easy identification. Otherwise, use bullets or other graphics. Parallel grammatical structure is used for all items in a list; for example, all items may be imperative statements, statements of facts, or quantities of articles. Also, all items should use the same tense (i.e., past, present, or future).

Examples

Not parallel: Use a fine file to create a smooth surface. [Imperative]
Painting the surface with matte paint will create a dull finish. [Future tense]
Shellacking the surface prolongs the life. [Present tense]

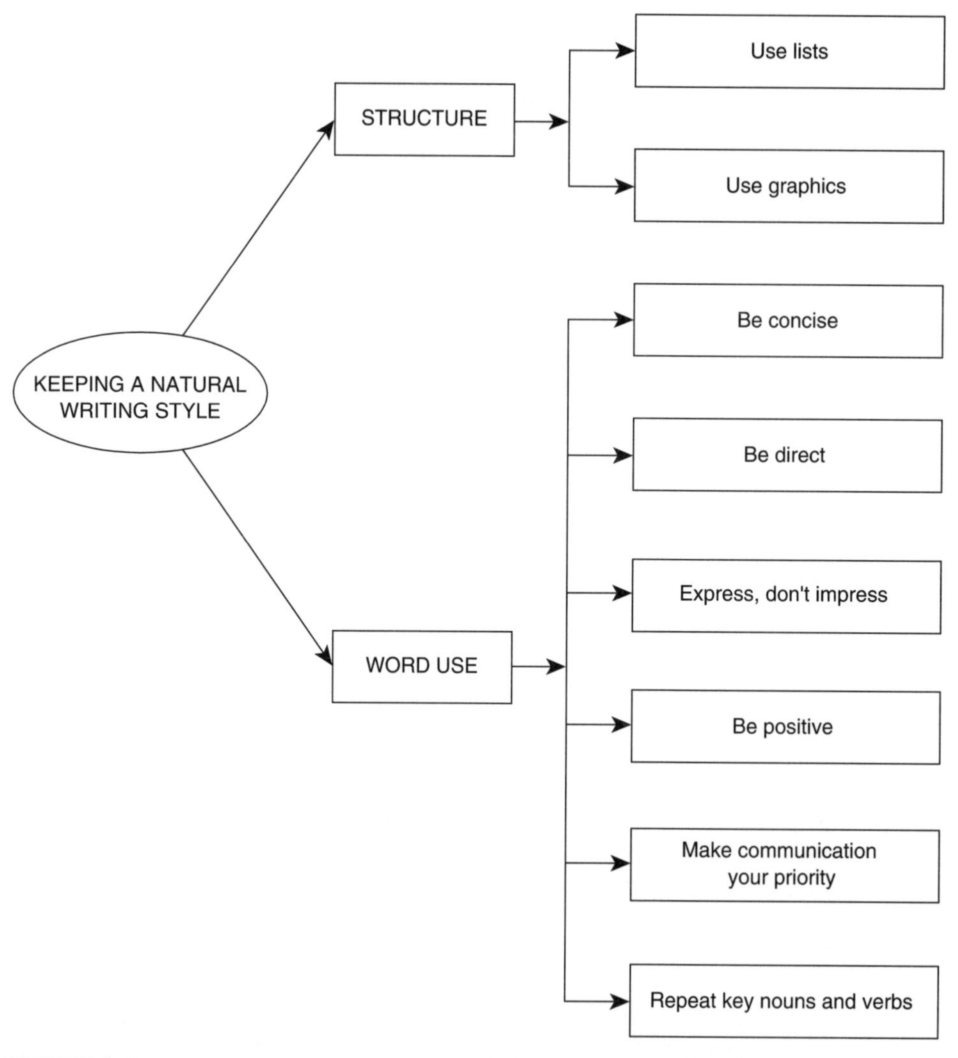

FIGURE 3–1
Keeping a Natural Writing Style

Parallel: [All imperative]
Use a fine file to create a smooth surface.
Paint the surface with matte paint to create a dull finish.
Shellac the surface to prolong the life.

Parallel: [All present tense]
Using a fine file creates a smooth surface.
Painting the surface with a matte paint creates a dull finish.
Shellacking the surface prolongs the life.

2. Use graphics. Use graphs, charts, tables, and drawings to simplify explanations and descriptions and to create mental images.

Word Use

1. Be concise in your choice of words. Carefully select the words or phrase to express your idea in as few words as possible.

Examples

Instead of writing *a large number of,* write *many.*
Instead of writing *in the course of,* write *during.*

2. State your ideas as directly as possible. Avoid circumventing the main points or adding unnecessary explanations to prevent confusing the reader.

Example

Indirect: Due to uncertainties in the weather, it is difficult to predict when the first flight will be. However, the preference is for tomorrow.

Direct: If the weather permits, the first flight will be tomorrow.

3. Write to express, not to impress. Pompous language can easily camouflage your intended meaning. Industry jargon and acronyms should not be used unless you are sure that your readers will understand their meanings.

Examples

Pompous Language
Instead of writing *contemplate,* write *consider* (e.g., "We will *consider* revising the design").
Instead of writing *endeavor,* write *attempt* or *try* (e.g., "The technician *attempted* to increase the test frequency").
Jargon
Instead of writing *facilitator,* write *administrative assistant.*
Instead of writing *RIF,* or *reduction in force,* write *layoff.*

4. Be positive in your information. Use of the word *no* or *not* may require judgment or interpretation of the meaning by readers. Rather, positively state what you mean.

Example

Negative: This acid is *not effective* for chemical etching.

Positive: This acid is *too weak* for chemical etching.

5. Make communication of your ideas your priority. Rules of grammar do not need to be rigidly followed when violation of these rules simplifies the understanding of your ideas. For example, sentences may end with a preposition, paragraphs may include only one sentence.

Example

Acceptable: Which supplier should the parts be shipped *to?*

6. Repeat key nouns and verbs whenever necessary. Changing nomenclature for the purpose of creating interest is potentially confusing to readers.

Example

Inconsistent: The surface of the *movable work platform* was usable, even though the wood needed to be repaired. . . . Therefore, salvaging the *scaffold* was feasible.

Clear: The work surface of the *scaffold* was usable, even though the wood needed to be repaired. . . . Therefore, salvaging the *scaffold* was feasible.

COMMUNICATING EFFECTIVELY

The following guidelines (see Figure 3–2) will help you to communicate effectively.

Contents

1. Use building block organization. Present your headings and information in a sequence that builds on the information previously presented and facilitates selective reading by readers not interested in reading the entire report. For example, in an experimental report, the sequence of headings may be as shown on page 4. This building block sequence for presentation of information begins with a statement of the purpose of the report, continues with the theory and procedure, and concludes with its findings.

2. State your purpose clearly. Make sure readers know what you hope to accomplish at the beginning of the document.

Example

The purpose of this report is to determine the relationship between blade pitch and air velocity.

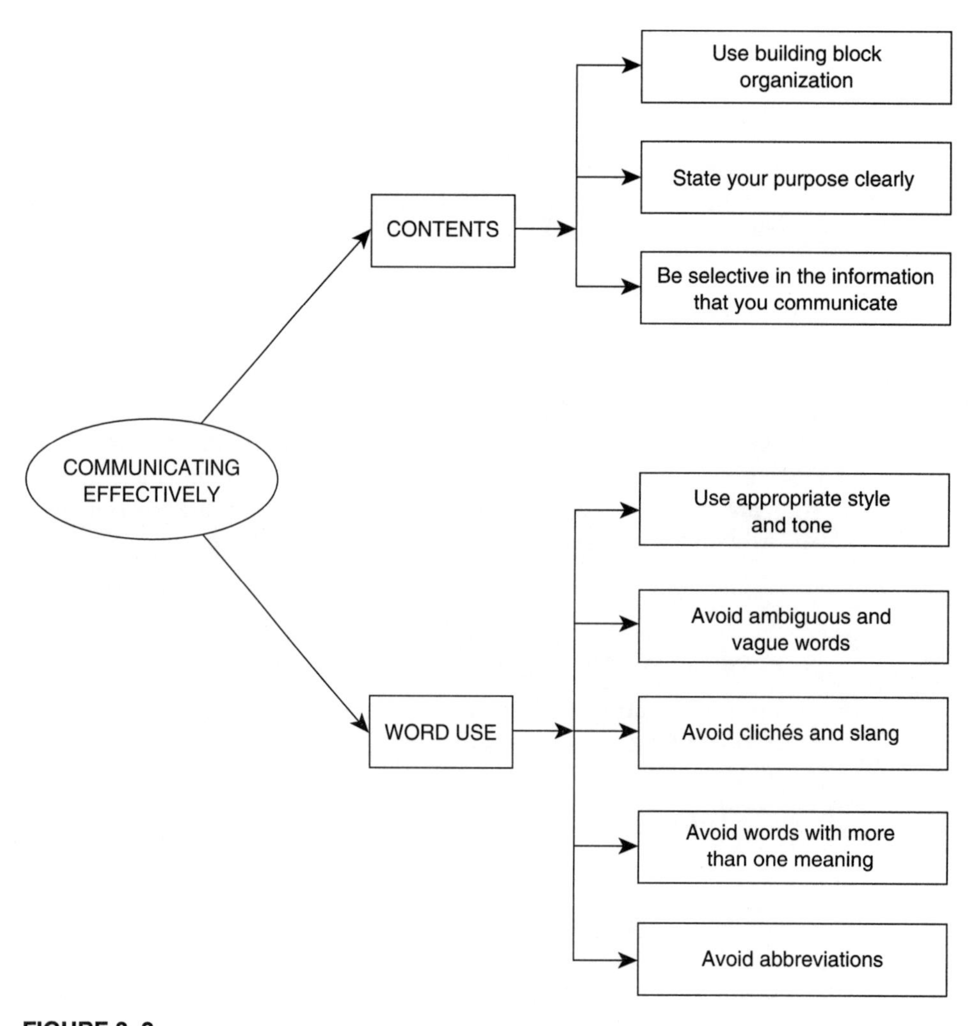

FIGURE 3–2
Communicating Effectively

3. Be selective in the information that you communicate. Tell the readers everything they need to know. However, be careful to include only relevant information that helps them to understand your message.

Word Use

1. Use appropriate style and tone. Theory should be instructional, instructions should be in the imperative, and conclusions should be affirmative. Demanding, vindictive, and derogatory tones cause the reader to become defensive and reluctant to objectively read and understand your message. Therefore, these tones are inappropriate for technical writing.

2. Avoid ambiguous and vague words. They create uncertainty and confusion for the reader.

Examples

Ambiguous: The flow of water was *affected.* (Was the flow increased or decreased?)

Clear: The flow of water was *decreased.*

Vague: *It* created a glossy appearance.

Clear: *The lacquer* created a glossy appearance.

3. Avoid clichés and slang. Clichés and slang give readers the impression that the writer is incapable of clear expression.

Examples

Avoid clichés such as *in any event, it goes without saying,* and *last, but not least.*

Avoid slang: Instead of writing *down the road,* write *later;*
instead of writing *cut corners,* write *reduce;* and
instead of writing *a drop in the bucket,* write *an insignificant amount.*

4. Avoid words with more than one meaning. They may be misinterpreted or unclear and require the reader to read further to understand their meaning.

Examples

Misinterpreted: *Since* the procedure did not yield the anticipated results, a new process was developed. (Was the procedure developed as a result of the unanticipated results, or after the unanticipated results?)

Clear: *Because* the procedure did not yield. . . . (Demonstrates cause and effect.) *After* a new procedure was developed. . . . (Demonstrates chronology.)

Misinterpreted: This is the *last* nozzle to be attached. (Was this the most recent or final nozzle to be attached?)

Clear: This is the *most recent* nozzle to be attached. (Demonstrates an ongoing process.) This is the *final* nozzle to be attached. (Demonstrates termination of the process.)

Unclear: *Once* we increased the strength, we had no additional failures. (Does *once* mean "only one time" or "after"?)

5. Avoid abbreviations unless the word is commonly abbreviated. Abbreviations can be misinterpreted by the reader.

Examples

Instead of writing *elect. engr.,* write *electrical engineering* or *electrical engineer* (whichever you mean).

However, *etc.* (for *et cetera*) is an acceptable abbreviation, but often is used in parentheses only. Outside parentheses, *etc.* is often replaced with a term such as "and so on."

WRITING CONVINCINGLY

Effective technical expression begins with the formation of coherent sentences and appropriate syntax (the combination of words and phrases to form sentences in a direct and natural manner). These elements (see Figure 3–3) are essential to communicate ideas and to persuade readers.

Sentences

1. Begin sentences with the central idea or concept. This focuses the reader's attention on this idea or concept. Any necessary explanations follow the central idea or concept.

Examples

Misleading beginning:	The increased density of the air means that the efficiency of the engine is increased.
Appropriate:	The efficiency of the engine increases with the increased density of the air.
Misleading beginning:	When the parts were strengthened, the number of reported failures decreased significantly.
Appropriate:	The number of reported failures decreased significantly when the parts were strengthened.

2. Break up long sentences. Concepts with more than one central thought are effectively communicated with shorter sentences.

Examples

Long:	A briefing concerning the use of our new Microstation computer program will be presented in the conference room next Tuesday at 2 p.m. for all students.
Appropriate:	A briefing concerning the use of our new Microstation computer program will be presented in the conference room next Tuesday at 2 p.m. All students must attend.
Long:	It is our philosophy that axial members of all structural systems have redundant load paths to minimize the possibility of structural failure and that the ends of these members have rigid joints to reduce deflection.
Appropriate:	It is our philosophy that axial members of all structural systems have redundant load paths to minimize the possibility of structural failure. Also, the ends of these members should have rigid joints to reduce deflection.

3. Avoid weak sentence beginnings. Unnecessary beginning phrases detract the reader's attention.

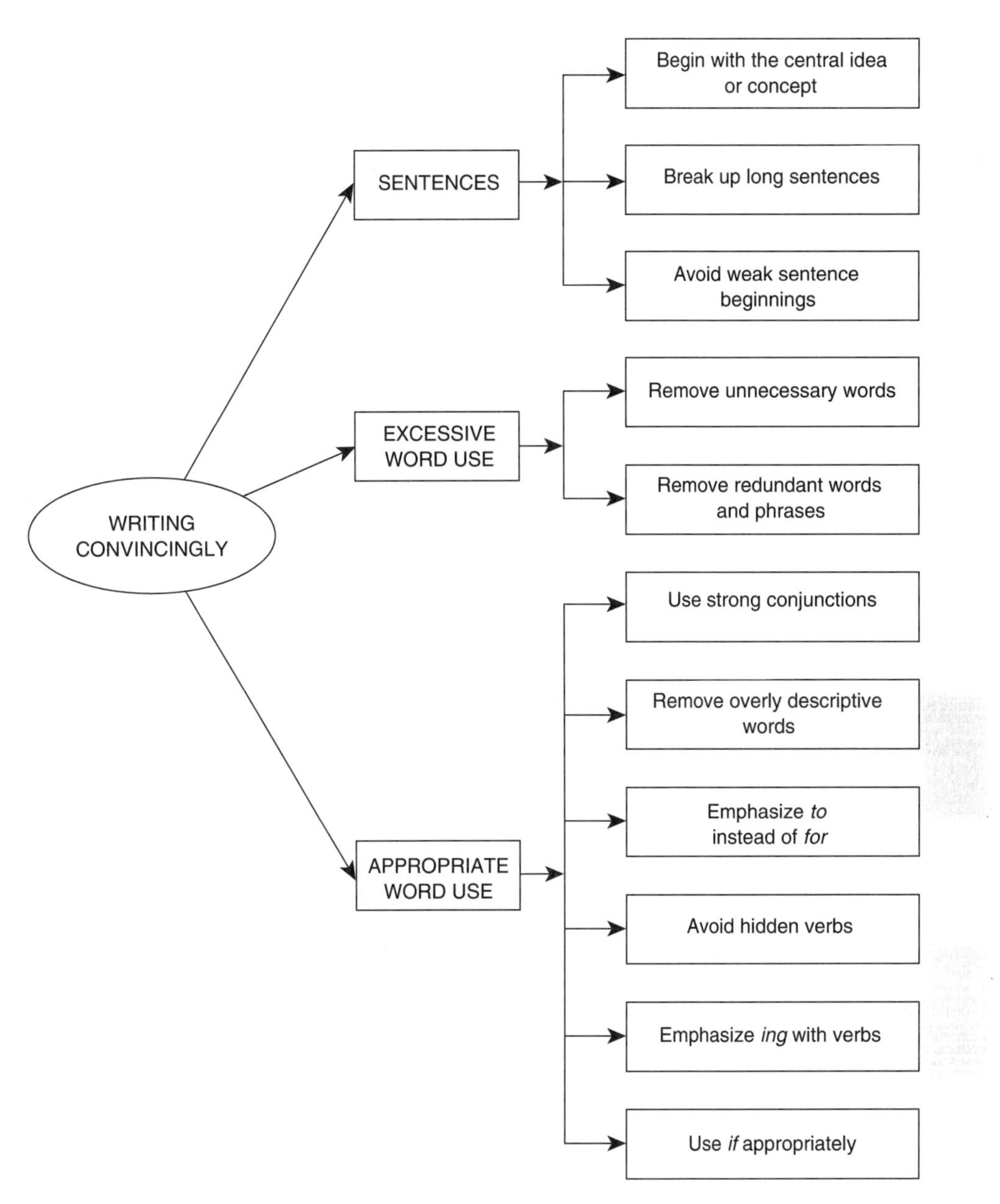

FIGURE 3–3
Writing Convincingly

Examples

Weak: *It was the* resilience of the material that prevented it from shattering.

Appropriate: *The* resilience of the material prevented it from shattering.

Weak: *As far as the rivets are concerned, they* were adequately strong.

Appropriate: *The rivets* were adequately strong.

Excessive Word Use

1. Remove unnecessary words. Remove words and phrases that do not clarify ideas or concepts.

Examples

Unnecessary: We purchased the computer *for the purpose of* increasing the efficiency of the laboratory.

Appropriate: We purchased the computer *to* increase the laboratory efficiency.

Unnecessary: The engine became overheated *because of the reason that* the fanbelt was not properly installed.

Appropriate: The engine overheated *because* the fanbelt was not properly installed.

2. Remove redundant words and phrases. When an idea is stated redundantly, it detracts from the remainder of the sentence.

Example

Redundant: The unlit, dark room was hazardous.

Appropriate: The unlit room was hazardous.

Appropriate Word Use

1. Use strong conjunctions. Conjunctions such as *however, but,* and *because* reveal interrelationships between connected thoughts to the readers. Weak conjunctions such as *and* conceal these relationships. Semicolons can be used with these conjunctions to separate interrelated thoughts that otherwise can be expressed independently.

Examples

Weak: The surface of the housing was corroded, *and* the steel was exposed.

Strong: The surface of the housing was corroded *because* the steel was exposed.

Weak: The testing machine was calibrated regularly, *and* the anvil was poorly maintained.

Strong: The testing machine was calibrated regularly; *however,* the anvil was poorly maintained.

2. Remove overly descriptive words. Words that are not simply understood and require interpretation detract from the intended meaning of ideas.

Examples

Overly descriptive: Each computer station is *equipped* with storage space.

Appropriate: Each computer station *includes* storage space.

Overly descriptive: To prevent corrosion, all equipment should be *enclosed* when not in use.

Appropriate: To prevent corrosion, all equipment should be *covered* when not in use.

3. Emphasize *to* instead of *for*. *To* followed by a verb makes your statement active.

Examples

Weak: The metallurgists have 2 weeks *for completion* of the research.

Strong: The metallurgists have 2 weeks *to complete* the research.

Also, in an imperative sentence that emphasizes cause and effect, placing the desired result (i.e., effect) in front of the cause further emphasizes this desired result.

Weak: Use dye penetrant *for detection* of surface cracks.

Strong: *To detect* surface cracks, use dye penetrant.

4. Avoid hidden verbs. Expressing the activity as an action verb emphasizes the activity rather than the actor of this activity.

Examples

Weak: The chromium plating *provides protection* against corrosion.

Strong: The chromium plating *protects* against corrosion.

Weak: This procedure *is applicable* to all groups.

Strong: This procedure *applies* to all groups.

5. Emphasize *ing* with verbs. Expressing the action after an article (e.g., *a, the*) detracts from the importance of this action.

Examples

Weak: *The removal* of the damper increased the flow of air.

Strong: *Removing* the damper increased the flow of air.

Weak: *The implementation* of the new test procedure will require the cooperation of all group members.

Strong: *Implementing* the new test procedure will require the cooperation of all group members.

6. Use *if* appropriately. Use *if* only for those actions that are uncertain to occur, not those actions whose time to occur is uncertain.

Examples

Inappropriate: *If* the ambient temperature is below 60°F, the rubber compound does not cure properly. (This action is certain to occur, only the time to occur is uncertain.)

Appropriate: *When* the ambient temperature is below 60°F, the rubber compound does not cure properly.

Appropriate: *If* the test is not completed this afternoon, we will complete it on Saturday. (It is uncertain that the test will be completed this afternoon.)

WRITING INSTRUCTIONS*

Write instructions as commands in the sequential order of performance.

- Each step of an operation or procedure usually begins with an imperative verb such as *measure, place,* or *cut.* Number the steps in the order of the operation. Specify any information necessary to perform a step or series of steps (e.g., a warning, or a time limitation) before that step or series of steps. Do not include statements that are informational only unless the information is required to understand the instruction that follows.

Examples

Informational only: *You may adjust* the speed of the ram, *if necessary.*

Imperative: *If necessary, adjust* the speed of the ram.

* Discussed in greater detail in Chapter 8.

Dangerous: The red lever shall be held in its lowest position while performing Step 6.2 *to prevent an explosion.*

Safe: *WARNING: To prevent an explosion,* hold the red lever in its lowest position while performing the following step (Step 6.2).

TENSE

Because the subject of your report is current, use the present tense to address any event that relates to the test. Also address other parts of the report in the present tense.

Examples

Incorrect tense: The environmental load test of the silicon computer chip *simulated* a 10-year life.

Appropriate: The environmental load test of the silicon computer chip *simulates* a 10-year life.

In the Introduction section of the report:

Incorrect tense: This report *will evaluate*. . . .

Appropriate: This report *evaluates*. . . .

In the Discussion of Results section of the report:

Incorrect tense: The experiment *confirmed*. . . .

Appropriate: The experiment *confirms*. . . .

Use the past tense for an event that was completed before the conception of the project and the report.

Example

Appropriate: This modification of the steel strut is to prevent future failures similar to the one that *occurred* last March.

Use the future tense for future or hypothetical events that are the topics of proposals. Using *would* and *should* for predictable outcomes is cumbersome. Rather, switch to the present tense.

Example

Cumbersome: If the modification of this laboratory is approved, student participation *would increase* by approximately 10 percent. An increase in student participation *would increase* maintenance expenses.

Appropriate: If the modification of this laboratory is approved, student participation *would increase* by approximately 10 percent. An increase in student participation *increases* maintenance expenses.

Table 3–1 summarizes all of the points discussed in this chapter. After writing your report, you can ensure your report's clarity by reviewing its conformance with these points.

KEY CONCEPTS

- Incorporating lists and graphics into your text, using a natural writing style, increases the readability of your writing.
- Selecting words and phrases that are not vague, or redundant, or have more than one meaning contributes to effectively communicating your message.

TABLE 3–1
Checklist for Reviewing the Clarity of Your Technical Reports

Keeping a Natural Writing Style

Structure

1. Are lists with headings used? Do these lists include introductory remarks?
2. Are visuals included when they would be helpful to the reader?

Word Use

1. Are your words concise?
2. Are your ideas stated directly?
3. Have you avoided pompous language?
4. Is your information stated positively rather than negatively?
5. Is communication of ideas your priority?
6. Are key nouns and verbs repeated when necessary?

Communicating Effectively

Contents

1. Did you build headings and information on ideas and concepts previously presented?
2. Did you state your purpose clearly?
3. Did you include only relevant information?
4. Were the abilities and limitations of your readers considered?

Word Use

1. Is the style and tone appropriate for this type of report?
2. Are ambiguous and vague words avoided?
3. Are clichés and slang avoided?
4. Are words with only one meaning included?
5. Are abbreviations avoided unless commonly abbreviated?

Writing Convincingly

Sentences

1. Do sentences begin with the central idea or concept?
2. Are sentences short enough for readers to understand easily?
3. Do sentences have strong beginnings?

Excessive Word Use

1. Have unnecessary words been eliminated?
2. Have redundant words and phrases been eliminated?

Appropriate Word Use

1. Are strong conjunctions used to interrelate thoughts?
2. Are overly descriptive words avoided?
3. Is *to* followed by a verb, emphasized rather than *for* followed by a noun ending in *ing?*
4. Are hidden verbs avoided?
5. Are actions emphasized with *ing* rather than introduced with an article followed by a noun ending in *al* or *tion?*
6. Is *if* used only for those actions that are uncertain to occur?

Writing Instructions

Do all instructions begin with an imperative verb?

Tense

Is the present tense used for all items discussed in the report, the past tense for all actions completed before the conception of the report, and the future tense for all future or hypothetical events?

STUDENT ASSIGNMENT

1. The following words can effectively communicate certain ideas, concepts, and information. However, they are sometimes used to impress when a simpler word is more appropriate. Substitute a simpler word for each of the following:

depict	utilize
facilitate	endure
magnitude	remote
retain	dramatic
equitable	impeccable

2. Revise the following sentences using the principles learned in this chapter to effectively communicate the writer's intent.
   - **a.** The group will decide the question of whether more office space will be needed.
   - **b.** Please indicate in your letter of transmittal the number of steel billets that you have a preference for.
   - **c.** The foreman together with his crew were responsible for start-up of the plant.
   - **d.** The flashing green light is indicative of the readiness of the vehicle to function.
   - **e.** Access ramps are geared for the handicapped.
   - **f.** It was only one tooth on the gear that did not comply with the tolerance requirements.
   - **g.** The steel bolts with the nuts have been properly plated.
   - **h.** The negligible bending stress in the member means that the required size of the member can be reduced.
   - **i.** The research group began the testing procedure.
   - **j.** The technicians have 5 weeks remaining for qualification of the components.
   - **k.** The handbook will be purchased by every engineer.
   - **l.** There are two groups that are responsible for the operation of the reentry vehicle.
   - **m.** The circuit board is intended to be built before the end of the year, while the project is not required to be completed for several years.
   - **n.** The videotape was about one hour long.
   - **o.** The size of the proposed plant is up in the air.
   - **p.** A meeting was held with our client for discussion of critical speeds.
   - **q.** The structural analysis has validity for the design.
   - **r.** The framers worked overtime after hours and did a first-class job.
   - **s.** The aircraft landing gear assembly tires need to be replaced immediately.
   - **t.** The client's request was complied with by our designers.
   - **u.** The bottom line is that the design of the existing facility will be a model for the new facility.
   - **v.** Use a micrometer for inspection of critical components.
   - **w.** Concerning the crane, it has sufficient reach to complete the job.
   - **x.** The catalyst helps speed up the chemical reaction.
   - **y.** To ensure proper operation, the tolerances of the components were reduced across the board.
   - **z.** This filter is not effective if the particle size exceeds 5 microns.
3. Revise the following instructions:
   - **a.** The load developed by the ram shall be increased in 5000-lb increments to 50,000 lb.
   - **b.** Place the soil sample in the oven to dry after compacting it in accordance with the procedure on page 57 of this manual.
   - **c.** Press the ON button. Make sure that your hands are away from the anvil before performing this operation.

CHAPTER 4

Rules of Practice for Lab Report Writing

The rules of practice for laboratory report writing allow writers to emphasize important data, phrases, and concepts in a manner that all readers understand; therefore, they are tools that writers use to achieve clarity. Because many of the rules of practice are inconsistent with each other, writers must consider the perspective, background, and knowledge of the readers in selecting the most effective rules to use. However, when these rules have been selected, they should be used consistently throughout the report.

This chapter provides general guidelines for use in writing reports. In your professional life, you will probably also consult the style manual produced for your specific discipline as well as any writing guidelines specified by your employer. You may also wish to consult a general style manual, such as *The Chicago Manual of Style* (14th Ed., Chicago, University of Chicago Press, 1993). This widely used style manual includes specific sections referring to scientific and technical writing.

ABBREVIATIONS AND ACRONYMS

The primary purpose of using abbreviations and acronyms (a word formed from the initial letters of a name, e.g., Department of Energy expressed as DOE) in technical writing is to simplify the reading, not the writing, of the text material. For example, if you referred to the lengthy title Nuclear Regulatory Commission only once or twice, it makes the most sense to write out the full name even though it is reasonable that most readers are familiar with the shortened form, NRC. However, when you intend to make the reference frequently, it should appear the first time as Nuclear Regulatory Commission (NRC), which alerts the reader that the shortened form, NRC, will appear later in the text.

On the contrary, when you are certain that all readers understand the meaning of an abbreviation or acronym, this abbreviation or acronym can be introduced into your writing in the shortened form without being defined. For example, the writer of an internal memo within the National Aeronautics and Space Administration may refer to this administration as NASA.

Use the following additional principles for abbreviations and acronyms.

General

- Include an abbreviations and acronyms section at the beginning of the report when you expect to use many different shortened forms. It should list all shortened forms in alphabetical order followed by the full words or names. The shortened form only can then appear in the text without further definition.

Example

ASTM	American Society for Testing and Materials
ERDA	Energy Research and Development Administration

- Begin an abbreviation with a capital letter only when the full word begins with a capital letter (e.g., K for degrees Kelvin but *vert.* for *vertical*).
- Do not begin a sentence with an abbreviation.
- When ending a sentence with an abbreviation, only one period is used. However, this may mislead the reader directly into the following sentence without a break in thought. Therefore, to avoid confusion, it is good practice to avoid abbreviations at the end of a sentence and either use the full word or revise the sentence to place the abbreviated word elsewhere.

Symbols

- Use the symbols # for *pound(s)* and % for *percent* in numerical analysis, tables, and figures. Also, for emphasis, the symbol % may be used in text when preceded by digits.

Do not write the symbol # to abbreviate *number* because, in technical reports, # is interpreted as *pounds.*

Examples

Instead of writing #10 *screw,* write *No. 10 screw.*

Units of Measurement

- When standard units of measurement are abbreviated, write them in the singular, and without a period. However, always use a period when abbreviating *inches* because it can be misread as the word *in* rather than *inches,* as intended (see Figure 4–1).

Examples

Wrong: 120 lbs of force
Correct: 120 lb of force
Wrong: 12.0 in long
Correct: 12.0 in. long

However, when units of measurement are not abbreviated, they are expressed in the plural when the number is greater than 1 (e.g., write 0.88 inch, but 1.02 inches).

Also, singular or plural units are determined by the number, rather than by the value of the number, specified before the units. For example, *0.75 kilogram* (which rep-

FIGURE 4–1
Abbreviation of Units of Measurement

- Singular and without a period except *in.* when abbreviating inches
- Only when preceded by digits

resents 750 grams) is singular because less than 1 kilogram is specified; and *50 centimeters* (which represents 0.50 meter) is plural because more than 1 centimeter is specified.

- Abbreviate standard units of measurement only when they are preceded by digits (e.g., write *35 gal* and *1800 rpm*).
- It is not necessary to define standard units of measurement when they are commonly used and understood (e.g., write *cu yd*).
- Use symbols for minutes (′) and seconds (″) of angle measurements only. Do not use these symbols to represent minutes and seconds of time or feet and inches of length. However, you can use ° as the symbol for degrees of angle measure or degrees of temperature (see Figure 4–2).

Examples

Angle:	171° 41′ 30″
Time:	15 m 45 s [or] 15 min 45 sec
Length:	10 ft 9 in.
Temperature:	72°

FIGURE 4–2
Measurement Symbols

Angles:	°, ′, ″
Time:	*h, m, s* or *hr, min, sec*
Length:	*ft, in.*
Temperature:	°

NUMBERS

Numbers can be expressed in technical reports as words (e.g., twenty-three) or digits (e.g., 23). Many of the common practices are inconsistent with each other. The following conventions (see Figure 4–3) are those typically used by most technical writers:

General

- Select a convention to use in the report. To avoid confusing the reader, you must be consistent in its use.

FIGURE 4–3
Number Expression Guidelines

Digits	Words
• Any measurement or data	• Any approximation
• Any number 10 or greater	• Any number less than 10 without units
• Any number less than 10 with units	• The beginning of a sentence
• Any number less than 1.0	
• When the numerical value of a quantity is important	

- Use digits for the following:

  Numbers derived by measurement or data (e.g., *2.5 ft*). Counted numbers (always expressed by integers) can be expressed either way (e.g., *twenty-three beams* or *23 beams*).

  Any number 10 or greater (e.g., *14 boilers*).

  Any number less than 10 when units of measurement are included (e.g., *8 ft/sec*).

  Any number less than 1.0. A zero is placed in front of the decimal point (e.g., *0.375*) unless it is not possible for the number to equal or exceed 1, as in probabilities (e.g., $p < .05$).

  For greater impact on the reader when the quantity is significant (e.g., *6 failures*). Most readers understand and remember digits more easily than words. This concept is especially useful in business correspondence.

- Use words for the following:

  Any number that is an approximation (e.g., *less than twenty-five hundred miles*).

  Any number less than 10 when units of measurement are not included.

  The beginning of a sentence when the sentence begins with a number.

  Example

  Wrong: 206 miles north of the city. . . .

  Correct: Two hundred and six miles north of the city. . . .

- In a series, use the same convention for all numbers, either digits or words. The convention is usually determined by the longest number in the series. (e.g., use digits to write *1 screwdriver, 1 wrench, 416 screws, 416 nuts, and 12 sheets of steel*, because 416 is greater than 10.)
- When two numbers are expressed together, express one in digits and the other in words to avoid confusing the reader. Usually, the shorter one is expressed in words. (e.g., *28 six-penny nails* and *twelve 3/16-inch screws.*)
- You may use engineering notation (mixed digits and words) for very large and very small numbers. (e.g., *38 thousand screws* and *12 microamps.*)
- Metric units

  It is sometimes necessary in a report to include units of measurement in the metric (SI [International System of Units]) system, as well as the English system. The convention in the United States is to include the metric equivalent in parentheses after the English units (e.g., *12.8 ft (3.90 m)*). For other industrialized countries where the metric system is used, include the English equivalent in parentheses after the metric units (e.g., *3.90 m (12.8 ft)*).

  When calculating the metric equivalent of English units, use the same number of significant digits in the metric equivalent as in the English units.

  Example

  Wrong: 346 gal (1.3096 m^3)

  Correct: 346 gal (1.31 m^3) [i.e., both English and metric include only three significant digits.]

Numeric Proper Nouns

Numeric proper nouns, common in technical writing, are those proper nouns represented by a number rather than a name such as Building 6 or Figure 2.

- Numeric proper nouns are always capitalized (e.g., *Table 10, Example 6.3,* or *Lane 1*). However, page numbers of literary works and reports are not capitalized because a page number is a sequential location rather than a proper noun (e.g., use *page 23* rather than *Page 23*) although chapter and paragraph numbers are capitalized (e.g., *Chapter 16* or *Paragraph 4.7.3*). It is not common for these numerics to be expressed in words (e.g., use *Printer 2* rather than *Printer Two*).
- Sizes of physical objects such as screws and sheet metal whose sizes are represented by nonrepresentative digits are capitalized (e.g., *No. 8 screw* and *11 Gage steel*) although the object itself is not capitalized (note *screw* and *steel* in the preceding examples).

EQUATIONS AND CALCULATIONS

Equations and calculations (herein referred to as equations) are frequently included in laboratory reports and engineering and design analyses. Use the following guidelines to present them in formal reports as an integral component of the text material:

- Using words, introduce an equation in the text before presenting the mathematical symbols and numerical calculations.
- Either indent the beginning of an equation from the left margin or center the equation on the line.
- Introduce an equation using mathematical symbols (i.e., variables, coefficients, and exponents) before numerically solving it.
- When showing the numerical solution to an equation, show the numerical calculations in equation form only and the final answer. Do not include intermediate arithmetic steps.
- When an equation is referenced later in the text, place an equation number or lowercase letter in parentheses inside the right-hand margin. The reference in the text to the equation number always appears subsequent to the appearance of the equation number.
- Leave one space before and after all operation signs (i.e., +, −, ×, ÷, and =). However, do not leave a space between the minus sign indicating a negative quantity and the quantity (e.g., write -8 *ft/sec*, not $- \ 8$ *ft/sec*).
- When several lines of equations are included, line up the equal signs. Therefore, the left margins may be irregular.
- When defining mathematical symbols used in an equation, include the units of measurement in parentheses after each definition. Use a list for defining more than one symbol.

Example

The combined compressive stress is defined by

$$f_c = \Sigma P/A + Mc/I \tag{6}$$

where:

f_c = combined compressive stress (lb/in.^{2})

ΣP = sum of the axial loads (lb)

A = area of the cross section (in.^{2})

M = bending moment about the x-axis (in.-lb)

c = distance to the compression fiber (in.)

I = moment of inertia about the x-axis (in.^{4})

Therefore,

f_c = 82,000/6.19 + 83,200(4.95)/106.3, and

f_c = 17,120 #/in.^{2}

- When an equation requires more than one line
  1. Separate the equation only at a plus (+) or a minus (−) sign. Do not separate terms that are multiplied, divided, or otherwise operated on. However, an expression within parentheses can be separated at a plus or minus within these parentheses.
  2. Begin the second and subsequent lines with the plus or minus sign aligned with the first symbol or number of that side of the equation directly above it.

Example

Wrong: F = 32.2(65.5 + 4.87)^2 + (45.0 + 21.9 − 6.0)[0.00452 +
2.00(52.9 + 79.9)$^{1.33}$]#

Correct: F = 32.2(65.5 + 4.87)^2 + (45.0 + 21.9 − 6.0)[0.00452
+ 2.00(52.9 + 79.9)$^{1.33}$]#

COMPOUND TERMS

Compound terms, or hyphenated words, help a writer clarify meaning and give emphasis that would otherwise be difficult to achieve.

Hyphenated words help a writer to clarify meaning by identifying the following:

- Compound adjectives modifying a noun: *double-acting pistons* and *two-way street*
- Compound nouns: *y-coordinate* and *passenger-miles*
- Compound verbs:

  The mechanic reverse-flushed the radiator.

  The technician heat-treated the steel.
- Closely related words in a phrase: *tool-and-die* and *black-and-white photograph*

Hyphenated words also help to avoid misinterpretation of phrases such as *heavy metal pollution,* which can be intended to be *heavy-metal pollution* (pollution by heavy metals), or *heavy metal-pollution* (heavy pollution by metals).

SPECIAL APPLICATIONS FOR COMMAS AND SEMICOLONS

Because technical writing includes content that is different from literary and journalistic writing, commas and semicolons are used for these special applications (see Figure 4–4):

- When only two elements are used in a series, a comma is not inserted between these two elements. However, when more than two elements are used in a series, a comma is usually inserted between the final two elements before the word *and* or before the word *or* for clarity. This convention is particularly helpful in reading technical writing because elements that appear in a series may include compound terms (e.g., *cause-and-effect* or *nuts and bolts*). This comma between the last two elements in a series that consists of more than two elements alerts the reader that the series is ending. This convention is used in technical literature published by the U.S. government and by most technical writers.

FIGURE 4–4
Special Applications for Commas and Semicolons

Commas	Semicolons
• Before the last element in a series	• To separate groups in a series
• In numbers greater than four digits	• To separate numbers in a series when any number includes a comma

Examples

Wrong: Each component is fabricated from either *steel, or* aluminum.

Correct: Each component is fabricated from either *steel or* aluminum.

Vague: Each participant must have one calculator, *pencil and paper and* one straight-edge.

Better: Each participant must have one calculator, *pencil and paper, and* one straight-edge.

- When a series contains groups that include individual elements, a semicolon is used to separate these groups.

Example

Confusing: Well-written technical reports usually include the following: correct punctuation, grammar, and syntax, mathematical accuracy of calculations, reasonable assumptions, logical conclusions, and feasible recommendations.

Improved: Well-written technical reports usually include the following: correct punctuation, grammar, and syntax*;* mathematical accuracy of calculations*;* reasonable assumptions*;* logical conclusions*;* and feasible recommendations.

- A comma is used to separate digits for numbers greater than four digits only. A comma is used in a four-digit number only when that four-digit number is included in a column or in a sentence that includes other numbers greater than four digits. Also, in a column of numbers, all decimal points line up with each other.

Examples

Instead of writing *4,647,* write *4647* in text.

But, write *73,987.*

Also, write

23,863.28
6,028.61
671.43
37,829.50
68,392.82

- When any number included in a series of numbers includes a comma, the numbers in that series are separated by semicolons.

Examples

76,287; 4876; 3650; 539; and 49,902

or

356, 82, 749, 271, and 86

- All other uses of commas and semicolons in technical writing are the same as for literary and journalistic writing.

KEY CONCEPTS

- The rules of practice for technical writing are your tools for emphasizing and clarifying the important technical concepts of your report.
- The rules of practice for technical writing promote clarity and allow you to emphasize your message easily.
- The perspective, knowledge, and background of the readers are considered in selecting the most appropriate rules and conventions to use in the report.

STUDENT ASSIGNMENT

Revise, if necessary, the following sentences to reflect the technical concepts related to numerical concepts, units of measurement, and equations:

1. The lifting was performed by four one-hundred ton cranes.
2. Several gal of gas are required for the trip.
3. The gap between the beams was 0.50 in..
4. The conveyor belt is 142 ft (43.28 m) long.
5. The requisition asks for twenty-seven machinists to be hired for the project.
6. The assembly is held together with 16 five-eighths-in. bolts.
7. We used 4 wire-mesh screens and two doorframes.
8. The bending moment at the center of the beam is 520 ft-lbs.
9. 50 microamps are required for the relay.
10. Our profit margin for the year was 18 percent.
11. The prototype of the Model MJ-2 helicopter achieved a maximum speed of 160 MPH.
12. The shop has 42 used lathes for sale.
13. It costs almost $3000 per week to run the operation.
14. The project is approximately 25% complete.
15. The Air Force will purchase four prototype airplanes for testing.
16. The eight cylinder engine was trouble-free.
17. Enclosed is your order for four hundred digital thermometers.
18. The half spent fuel cell can be dangerous.
19. Use a #8 sheetmetal screw for Step 7.
20. We completed 16 thousand hours without an accident.
21. Thirty-two in/sec is too fast to maintain control.
22. The instructions were in an easy to-read format.
23. The natural frequency of the system is 38 hz.

24. We purchased 100 pounds of sulfuric acid.

25. The slab is .75 in. thick.

26. Two 40-pound bags were adequate to resurface Lane 1.

27. $M_2 = aP(1 - 2/\pi)$ (Eq. c)

28. The height of the wall is 96 in from the floor to the ceiling.

29. The motor speed is 3500 revolutions per minute (rpms).

30. The list of equipment includes two tractors, one backhoe, one grader and, 28 hand shovels.

31. The dry weight of the vehicle is 8240#.

32. The size of the housing, 12′ 6″ × 9′, is too large to fit in the enclosure.

33. The warm environment design temperature is 150° Fahrenheit.

34. 6,400 gal of fuel were consumed.

SECTION II

Elements of the Lab Report

Section II presents the typical elements of laboratory reports: the cover materials, the graphics, and the report sections. The report sections discussed include samples which use different style formats to show the different visual effects on readers.

A completed laboratory report, at the discretion of the instructor, may include most of the sample report sections shown in the following chapters. Although these report sections are discussed independently, in Chapter 10 they are integrated with each other to comprise a coherent report.

Chapters 6 through 9 represent the order these sections and materials are usually included in reports. However, follow the instructions for order of presentation given by your instructor.

The sample sections of laboratory reports include critiques of the *structure* and *content* of the writing.

Structure includes

- Format and organization
- Clarity and conciseness of expression
- Use of graphics
- Syntax
- Parallelism of expression
- Consistency and conformance to standard
- Practices, persuasiveness
- Tone
- Tense
- Readability

Content includes

- Sense of audience
- Understanding of purpose
- Addressing the appropriate issues

The exercises at the end of each chapter include student samples for you to critique to give you knowledge of these sections, and sections of reports for you to write to give you experience.

CHAPTER 5

Graphics

Graphics (tables, graphs, charts, drawings, diagrams) in reports present in a visually attractive and instantly comprehendible manner what the written text presents in a linear or sequential manner. Graphics make information easier for readers to understand and remember and can be used to help them interpret concepts and ideas. (See Figure 5–1.)

For many readers, graphics are the primary source of information. These readers read the written text only when they do not understand the graphics. For this reason, clear graphics to communicate your message are important.

GUIDELINES FOR USING GRAPHICS

An effective graphic presents limited information. Only one idea or concept is emphasized for a clear message. A graphic should be large enough to be read easily. However, you may use as many graphics as necessary to communicate your information.

Consider the needs of readers when selecting graphics.

- Tables and charts with data and facts communicate information to others.
- Graphs demonstrate the relationships between variables.
- Diagrams demonstrate technical procedures and operations.
- Drawings demonstrate the shape and size of objects.

Graphics assist writers to communicate ideas and concepts. These graphics should be integrated with explanations to visually demonstrate key points in the text. For example, the following statements can be incorporated into the text:

> The voltage spike seen at t = 12 seconds in Figure 4 was caused by a surge in the line current.

or

> A study of the critical path diagram (see Figure 2) will help clarify the reason the analysis of the aft bulkhead was not completed by the scheduled date.

Either of these statements would require extensive description without graphics. Also, the graphics minimize the possibility of misinterpreting the message of the writer.

Please note that all graphics shown in this text are for illustrative purposes only and the information included in these graphics is not to be used in any technical context.

FIGURE 5–1
Types of Graphics

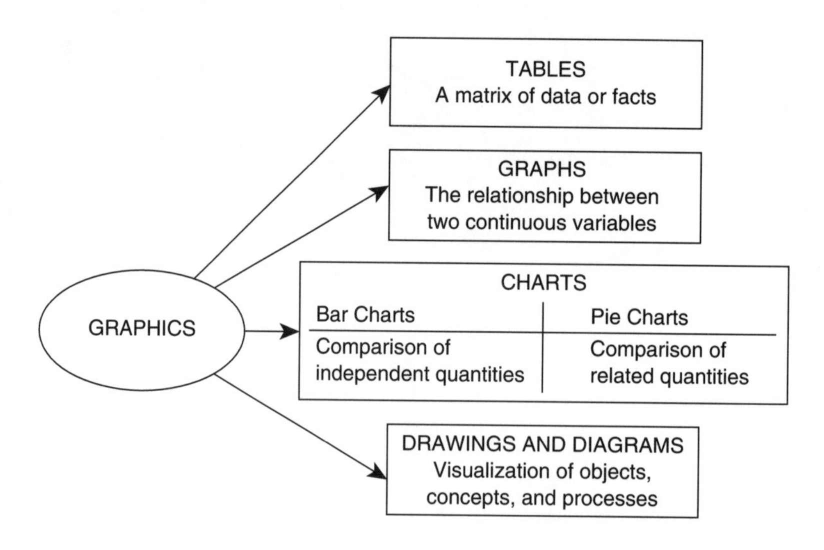

Referencing Graphics

Graphics are *referenced sequentially* in the text of the report; that is, the reference to Figure 1 always precedes the reference to Figure 2. They are placed in the report in the same order they are referenced.

The sets of figure numbers and table numbers are each *numbered sequentially* beginning with 1 (e.g., *Figure 1, Table 1*), and usually include captions (a brief description of the contents of the graphic).

All graphics in a report are labeled *figures*, except for tables, which are labeled *tables*. However, when a report includes only several figures and tables, all graphics are usually labeled *figures* for the reader's convenience.

Locating Graphics

When possible, a graphic appears on the same page as its reference in the text. However, the graphic never precedes its first reference in the text. When the graphic appears on a page other than where it is referenced, the page number is usually included with the reference.

Reports that are only several pages long frequently place all graphics with captions at the end of the report. Also, many reports include graphics pertaining to a section at the end of that section.

Labeling Graphics

Figure numbers with captions appear below the graphics except when the figure number with caption clutters the data or scale along the horizontal axis. Then, the figure number with caption may be placed in the upper half of the graphic field. Table numbers with captions appear near the top of the graphic field. Any facts pertinent to the understanding of, and any limitations on the validity of, the information in a graphic are clearly indicated in the caption or graphic field.

Developing Graphics with Computers

Many students develop their own graphics with computers. Programs are available for personal computers that make it possible for you to design professional-looking graphics to be inserted into your reports. Because of the capability, speed, and low cost of these programs, graphics can easily be generated and revised until the graphic representation is suitable for your purposes and layout.

These programs ordinarily allow you to select the desired type of graphics (e.g., table, graph), enter the data, and add relevant words, numbers, and symbols. These program capabilities make it easy to integrate these graphics within the text of your report.

Spreadsheets (computer programs that perform repetitious calculations and print the data and results in tabular form) save time, minimize the probability of mathematical errors, and add a professional appearance to your graphics not easily attainable otherwise. Graphing programs can accurately plot the closest approximation of a smooth curve (see Step 4 in "Labeling the Scales" later in this chapter) through a set of data points, thereby eliminating the writer's estimate.

TABLES

Tables are often used because the information may be presented without interpretation (analysis or calculation) by the writer. However, numerical data are frequently interpreted by the writer and presented in the same or subsequent tables. Tables present data or facts in matrix form (a rectangular array of numerical quantities or facts) and give the reader latitude to analyze and understand their meaning. For example, data collected from a laboratory experiment are ordinarily recorded in tables (see Figure 5–2). Also, tables are used effectively for non-numerical facts (see Figure 5–3). Tables may be accompanied by graphs or charts that demonstrate the relationship of the data or facts.

Follow these guidelines to set up a table:

- Place the table number with a title or caption near the top of the table.
- Label the top of each column and the left side of each row with a title.
- Separate column and row headings from data with a double or bold line.
- When all units of measurement for the table are the same, place the units with the caption of the table (e.g., "Length, in."). Otherwise, include the units of measurement in parentheses with the column or row headings.
- For a table with numerical data, line up the decimal points if used, or the last digit in each column.
- Indicate detailed explanations for data or facts in the body of the table with a superscripted number after the data or facts in the table. The explanations appear below the table (see Figure 5–2).
- When tables have more than five columns and five rows, when possible, separate the data or facts into subsets to encourage reader attention. Then, either display the separation of these subsets with double or bold lines, or display these subsets in two or more tables.

Critique of the Sample Tables

Figure 5–2: Figure 5–2, a six-column by five-row table from a student laboratory report, is divided into three subsets separated by double vertical lines. The left section, "Wheel Speed," is the independent measured variable. The center section, which includes "Brake Load" and "P_{NET}," are the dependent measured variables. The right section includes three columns of calculated data (results). Although the methods for determining the calculated data are appropriately not indicated, sample calculations for determining these data should be included in a prior section of the report.

FIGURE 5–2
A Table From a Student Laboratory Report

TABLE 6
Turbine Test Data and Results—Nozzle Setting = 6

Wheel Speed, N	Brake Load	P_{NET}	$N \times P_{NET}$	Brake Horsepower	Efficiency η
(rpm)	(lb)	(lb)	(lb/min)		(%)
300	8.2	6.4	1920	0.55	52
400	7.3	5.4	2160	0.62	58
500	5.7	4.0	2000	0.57	54
600	4.5	2.8	1680	0.48	45
670[1]	3.0	1.3[2]	870	0.25	23

[1] Maximum wheel speed.
[2] P_{NET} less than 2.0 is estimated.

FIGURE 5–3
A Non-Numerical Table Showing Uses of Stainless Steel

TABLE 4
Types, Composition, and Properties of Stainless Steel

Type No.	*Composition*	*Properties*
302	Basic Type Cr 18%, Ni 8%	Good formability.
304	Lower C	More weldable than 302.
316	Higher Mo	Resists salt water.
317	Higher Mo than 316	Good heat resistance, excellent corrosion resistance.
405	Al added to Cr 12%	Excellent heat resistance.

A title at the top of the table identifies the recorded data. Specific information concerning the nozzle setting is included with the title.

Each column of the table is labeled, and the units of measurement are indicated in parentheses. Each row is identified by the selected wheel speed for which the data are measured. The row headings are separated from data by a double line.

The decimal points or last digit for the data in each column are lined up. All data are included with the appropriate number of significant digits.

Data in the table that require additional explanations are superscripted in the body and are explained at the bottom of the table.

Figure 5–3: Figure 5–3, a non-numerical table used for determining the composition and description of different types of stainless steel, is divided into three columns. Units of measurement, not applicable for non-numerical data, are not included. Each row is headed with a type of stainless steel. A bold line separates column headings from data. A title at the top of the table identifies its contents.

GRAPHS

A graph is a visual interpretation of data that shows the interrelationship between two continuous variables such as voltage and time, or stress and strain. Graphs are used for the following reasons:

- A table in a laboratory report is ordinarily accompanied by a graph that includes the experimental data points and a smooth curve approximating the path of these points to help the writer determine the relationship between the two variables studied in the laboratory. The scatter of these data points on either side of this smooth curve is a measure of the validity of the data.
- A theoretical or design graph (a smooth graph without experimental data points) helps the reader understand the relationship between the variables or is a source of technical information for design or analysis.

Use the following guidelines for drawing a graph. (Note: The use of a graphing program eliminates the need for drawing your graph. However, the responsibility for following these guidelines belongs to the writer, not the computer.)

Setting Up the Page

1. Use standard engineering graph (coordinate) paper, for example, 10 lines to the inch, for graphs. Lines spaced 1 inch apart (main divisions) should be darker than intermediate lines (divisions). This paper is available in any engineering or drafting supply store. Graphs are usually sufficiently large so that they fill at least one-half of the page. Frequently, graphs fill the entire page.
2. Plot the independent variable (the variable that is controlled during a test or selected) along the horizontal (x) axis and the dependent variable (the variable that is measured during a test or determined) along the vertical (y) axis, except when this is contrary to standard practice. Usually, the axis of the independent variable is longer than the axis for the dependent variable.
3. Allow space below the horizontal axis, and to the left of the vertical axis, for numbering the main divisions and titling the scale of each axis.
4. Darken the horizontal and vertical axes to make them prominent.
5. Place the title (figure number and caption) of the graph below the graphic field. However, when this clutters the data or scale along the horizontal axis, place the title above the top center of the graphic field.

   The caption of the graph includes a statement of its contents or its variables (e.g., "Temperature vs. Deformation"). When the caption is a statement of the graph's variables, the first variable in the statement is usually the independent variable (e.g., temperature is the independent variable in "Temperature vs. Deformation").

   When the curve of only one specimen is shown in the graph, indicate the specimen's material and size (e.g., "2024-T4 Aluminum, 1/2-in. × 2-in. bar") with the caption. When the graph is titled below the graphic field, include this information with the statement of the variables. When the graph is titled in the graphic area, include this information directly below the statement of the variables.

Selecting the Scales

1. Select a scale large enough so that significant changes in the curve are apparent but small enough so that laboratory errors are not magnified.
2. Select scale units along the main divisions of the graph that are divisible by 10 for 10 lines to the inch coordinate paper and divisible by 4 for 4 lines to the inch

coordinate paper so that the intermediate divisions can be readily interpolated. For example:

For 10 lines to the inch coordinate paper, use main division scale units of 1, 2, 5, 10, or 20; the intermediate divisions are, respectively, 0.1, 0.2, 0.5, 1.0, or 2.0.

For 4 lines to the inch coordinate paper, use main division scale units of 1, 2, 4, 10, or 20; the intermediate divisions are, respectively, 0.25, 0.5, 1.0, 2.5, or 5.0.

This may require using less than the entire length available along the axis.

3. Number the graph 0,0 at the origin, and increase the numbers to the right, and up, respectively. When the first data point is considerably greater than zero, a section of the corresponding axis can be cut with a section-cut symbol (e.g., ≀≀). The plot of the curve should fill as much of the graph area as possible (also see Step 2).

Labeling the Scales

1. For the horizontal axis, place the title of the variable and the numbers so they can be read from the bottom of the graph. For the vertical axis, place the title of the variable so it can be read from the right-hand side and place the numbers so they can be read from the bottom of the graph.

Number the axes at the main divisions of each scale only.

Include the name of the variable measured and the units of measure, if any, in each axis title. The symbol for the variable is sometimes included. The units of measure can be placed in parentheses. For example, an axis title may read "Moment, M (ft-lb)."

2. The number of significant digits shown at the main divisions of each axis should be consistent with the precision of the measuring instruments or reliability of the results. When scientific notation is used, include the base 10 with its exponent only with the last (and therefore largest) numbers shown along the axis. Do not include the base 10 with its exponent with the title of the axis (when you include scientific notation with the title of the axis, it is unclear whether the number read from the scale has *already been* multiplied by the power of 10 or *needs to be* multiplied by the power of 10).

3. Show data points on the graph using a 1/10-inch diameter circle with a point in the center. When multiple specimens are shown on one graph, use a different symbol with a point in the center for each specimen; for example, use a triangle or square. Use a template for drawing these symbols.

For a theoretical or a design curve drawn from points calculated by use of an equation, show the points on the graph with a point rather than a symbol.

4. When the theoretical or design curve of the engineering variables is smooth (as is typical for two continuous variables), the curve representing the laboratory specimen is also shown smooth.

The curve will probably not be able to pass through all data points; therefore, select a curve that approximates, as closely as possible, the straight segments of lines connecting the points. An equal number of points should fall above and below the completed curve. The precise method of doing this is the Method of Least Squares (see any statistics text).

The completed curve may touch the edges of the symbols but should not pass through them.

For a theoretical or design curve, draw the curve passing through all the points.

Use a French-curve to draw all continuous curves.

5. For multiple specimens shown on one graph:

   Use a different line convention for each curve—solid, dash, dot-dash. Do not use a different color for each curve because colors do not reproduce when photocopied.

   Label each curve so that it reads horizontally. Use a two-section arrow to connect each label with each curve. The pointed section of the arrow should be perpendicular to the curve at the point of contact; the tail section of the arrow should be a short horizontal line drawn at mid-height from the label, and it should begin at either the front or rear of the label. A legend may be included in lieu of these labels with arrows to identify the line conventions.

See Figure 5–4 for a multiple specimen curve and Figure 5–5 for a design curve.

Critique of the Sample Graphs

Both figures are shown on rectangular coordinate graph paper; however, Figure 5–4 is shown on semilog graph paper, which may be used when the variation of the dependent variable is significantly more pronounced at the lower values of the independent variable, and these variations are to be emphasized.

In the design curve of Figure 5–5, the independent variable—the variable known by the user for determining the other variable—follows standard practice and is shown along the horizontal axis. However, contrary to standard practice, in Figure 5–4, the independent variable is shown along the vertical axis. When stress (or load) is one of the variables, it is the customary practice to show it along the vertical axis. (For the purposes of this text, Figures 5–4 and 5–5 are shown smaller than one-half page.)

The horizontal and vertical axes of both figures are displayed more prominently than any division lines of the graph. The horizontal scale of Figure 5–4 does not begin at zero; however, this is inherent when using semilog graph paper, and therefore a section-cut symbol, as indicated in Step 3 of "Selecting the Scales," is not included. Also, the main divisions of each scale are numbered, and its lines are more prominent than the intermediate lines.

The scales are sufficiently large for the curves to be read and interpolated easily. In Figure 5–4, which includes experimental data points, the scale is not so large for the scatter of these points to distort an otherwise smooth curve. The main divisions of Figures 5–4 and 5–5 are divisible by 10.

The figure titles below the figures include a statement of its contents. Figure 5–5 also includes specific information concerning the applicable value of H/D.

FIGURE 5–4
A Multiple Specimen Curve

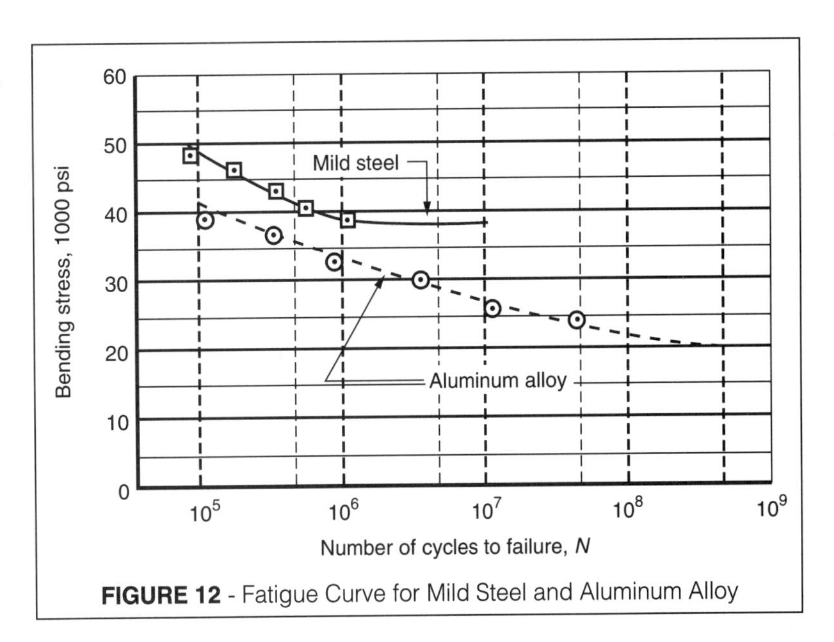

FIGURE 12 - Fatigue Curve for Mild Steel and Aluminum Alloy

FIGURE 5–5
A Design Curve

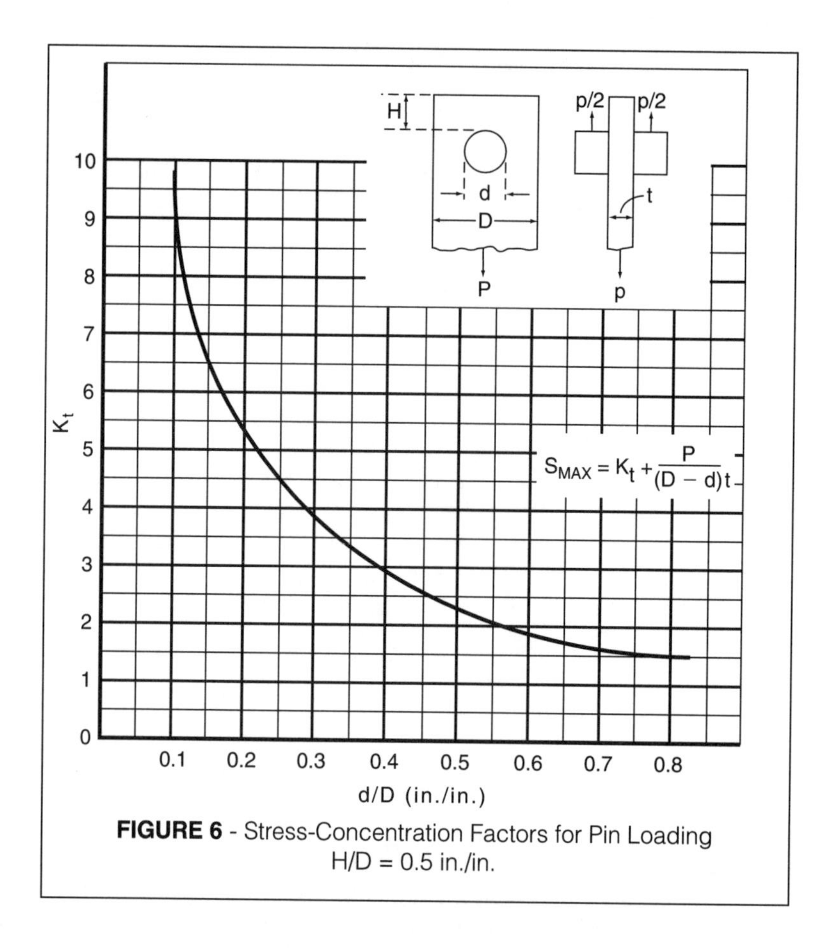

FIGURE 6 - Stress-Concentration Factors for Pin Loading
H/D = 0.5 in./in.

All numbers and the titles of the variables represented by the horizontal axes can be read from the bottom of the graph. The titles of the variables represented by the vertical axes can be read from the right-hand side of the graph. The titles of all variables are descriptive and include units of measure, when applicable. The number of significant digits at the divisions of the scales is consistent with the precision of the instrumentation or reliability of the results.

In Figure 5–4, an experimental multiple specimen curve, data points are shown with points enclosed with a different symbol for each specimen, and a different line convention is used for each curve. The curves are drawn smoothly with an equal number of data points above and below each curve. These curves touch the edges of the symbols but do not pass through them. Each curve is labeled horizontally and uses a two-section arrow, which is perpendicular to the curve at the point of contact.

In Figure 5–5, a design curve without symbols, the smooth curve passes through all points. Figure 5–5 also includes a sketch demonstrating the procedure to use the curve and to apply the data obtained from it.

CHARTS

Charts are very effective for comparing quantities because they are easy for readers to understand without interpretation. There are two basic types of charts. Bar charts compare independent quantities such as time spent for design versus time spent for manufacturing. Pie charts compare dependent quantities that represent parts of a whole, such as the percentage of your inventory of parts used for each project.

FIGURE 5–6
A Sample Bar Chart Showing a Time-Dependent Phenomenon

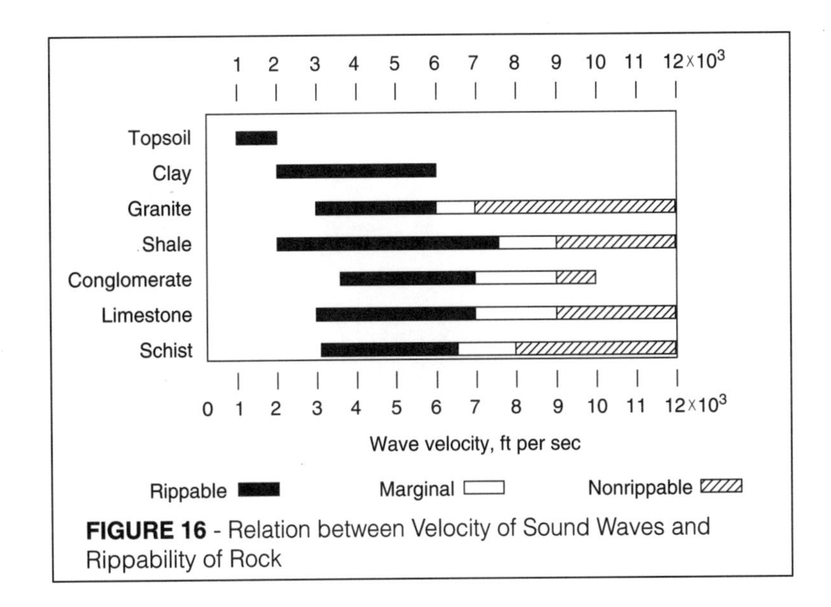

FIGURE 16 - Relation between Velocity of Sound Waves and Rippability of Rock

Bar Charts

Bar charts show sets of horizontal or vertical parallel bars drawn to scale to compare independent quantities where each quantity represents one variation of the same set. For example, a bar chart can compare the expected lives of different automobiles for similar driving conditions.

Most writers prefer vertical (rather than horizontal) bars in bar charts, especially when the compared quantities are counted quantities, such as production of units for each model or dollars spent by each department. However, horizontal bars are used when the quantities being compared are time, such as projected time for completion of different phases of a project, or horizontal phenomena, such as stopping distance for varying road conditions.

For emphasis, the bars in a bar chart are usually shaded, cross-hatched, or otherwise marked. Bar charts sometimes show multiple sets of adjacent bars when the bars within each set are interrelated. For example, the average ambient seasonal temperatures for a period of 5 years can be shown with five sets of four adjacent bars, each bar of a set representing one of the four seasons. The marks for each bar of the set differ from the other bars in the set, but the markings in all five sets are the same. A key (sometimes called a legend) demonstrates the designation for the markings on each of the adjacent bars.

The axis representing quantity is clearly labeled and includes the units of measurement, if any. Bars are identified by labels at their bases and, sometimes, with a key when they are dissimilarly marked. It is helpful to the reader to indicate above the bar the numerical quantity this bar represents. To avoid misinterpreting the information in the chart, bars should always begin at zero (frequently, bar charts used for advertising a product begin at other than zero to exaggerate differences).

See Figures 5–6 and 5–7 for sample bar charts.

Critique of the Sample Bar Charts

Figure 5–6, a horizontal bar chart, compares the rippability of rock by measuring the velocity of sound waves—a horizontal field measurement—through these rocks. The figure, with the number and caption at the bottom, includes a key for the dissimilar markings of each bar, which identify the ease of rippability at different wave velocities.

The vertical axis of Figure 5–6 includes the names of various types of rock to be compared. The horizontal axis is titled and includes the units of measurement. It begins at

FIGURE 5–7
A Sample Bar Chart Showing Multiple Sets of Related Quantities Phenomenon

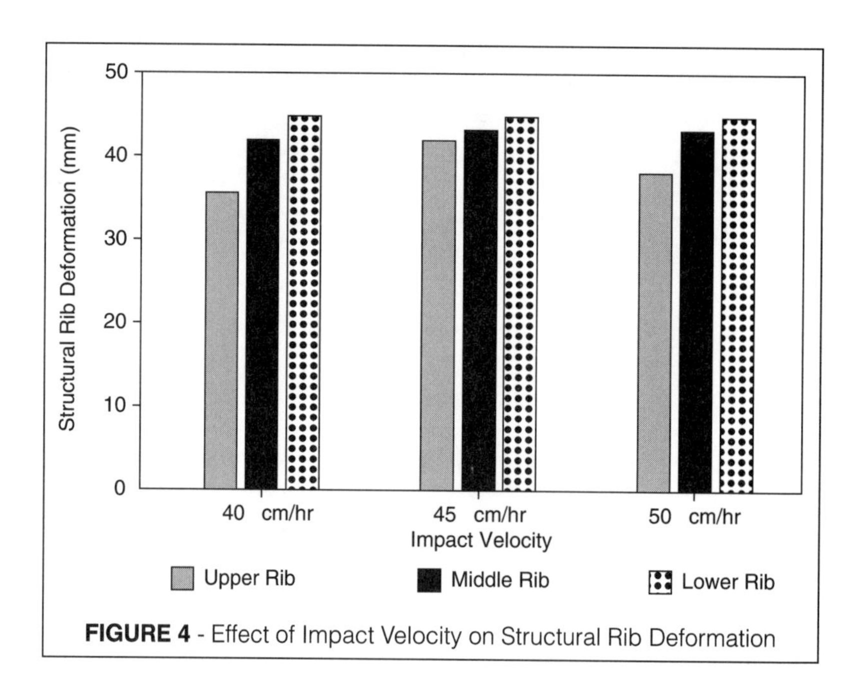

FIGURE 4 - Effect of Impact Velocity on Structural Rib Deformation

zero and is numbered on the top as well as the bottom for ease of reading. The last number along the horizontal axis includes scientific notation that clearly instructs the user to multiply all numbers by 10^3. All numbers and titles can be read from the bottom.

Figure 5–7, a vertical bar chart with multiple sets of adjacent bars, compares sets of upper, middle, and lower structural rib deformations at different impact velocities. A key is included to differentiate the different ribs at the same velocity. All numbers, the units of measurement, and the title of the horizontal axis can be read from the bottom. The vertical axis includes units and can be read from the right-hand side of the page.

Pie Charts

Pie charts show circles cut into pie-shaped wedges representing parts of a whole. Each piece of the wedge is individually shaded, cross-hatched, or otherwise marked to distinguish it from the others and can be lifted away (exploded) from the remainder of the pie for emphasis.

Each wedge is identified or labeled immediately outside the wedge and has an arrow pointing to the field. When the field is sufficiently large, the identification or label can be displayed within the field. Wedges should be arranged in sequence by size, largest to smallest, clockwise, beginning at the 12-o'clock position.

The numerical quantity and percentage of the whole represented by each wedge is shown directly below the identification. All identifications are printed horizontally. When they are placed outside the wedge, the tail of the arrow begins with a short horizontal segment at the mid-height of the beginning or end of the identification, and then continues in the radial direction approximately halfway into the wedge. The arrow terminates at a section-cut symbol. Some writers prefer to terminate the arrow at the perimeter of the wedge rather than halfway into it.

See Figure 5–8 for a sample pie chart.

Critique of the Sample Pie Chart

Figure 5–8, a pie chart, compares the relative expenditures of the operation and maintenance costs for equipment for a given year. A title appears at the bottom of the figure to identify its contents.

FIGURE 5–8
A Pie Chart

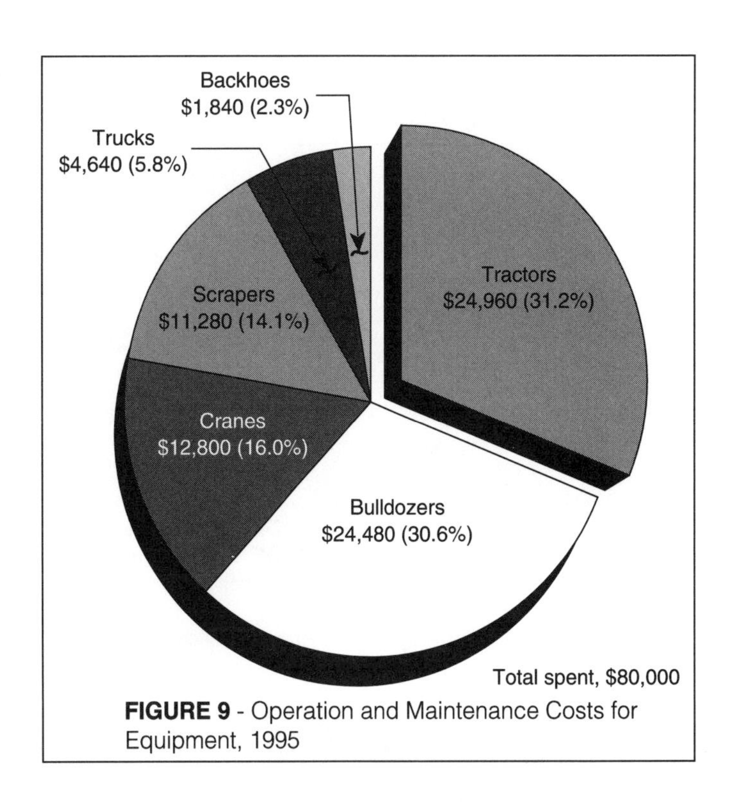

FIGURE 9 - Operation and Maintenance Costs for Equipment, 1995

Each wedge is identified with a title. The cost and percentage of the whole are indicated directly below the title. This information is included within the wedge when space is adequate. Otherwise, this information is placed outside the wedge and has a two-segment radial arrow terminating at a section-cut approximately halfway into the wedge.

Wedges are arranged clockwise by size with the largest wedge beginning at the 12-o'clock position. One wedge, "Tractors," is lifted away for emphasis.

Each wedge of the pie is marked differently from the others. Because each wedge is identified within the diagram, a legend is not included.

DRAWINGS AND DIAGRAMS

Drawings are used in reports to help the reader visualize physical objects such as pumps and buildings. They can emphasize the following:

- The interior of objects, with section views through critical planes and hidden lines
- The assembly of components, including exploded views
- Critical elements by eliminating noncritical elements
- Material, with cross-section symbols

Diagrams are also used in reports to help the reader visualize linearly dependent concepts and processes such as schematic diagrams that demonstrate how an electronic system works. (Organization charts showing management structure, and flowcharts showing computer algorithms are actually diagrams, because their elements show linear dependency.) Diagrams should be simple but contain the details that the readers need. They should be large enough to read easily.

A desktop computer can provide professional-quality drawings and diagrams at little expense.

See Figure 5–9 for a sample drawing and Figure 5–10 for a sample diagram.

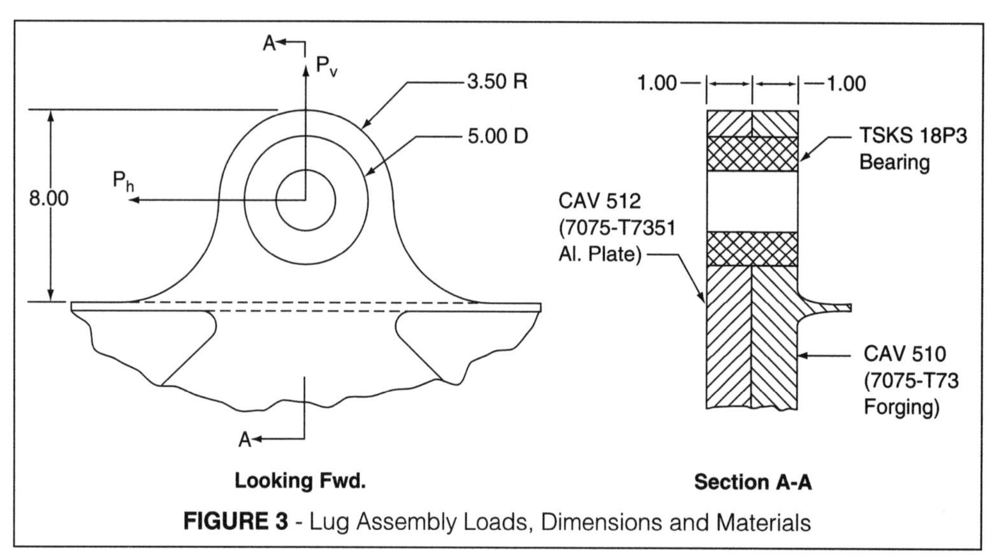

FIGURE 3 - Lug Assembly Loads, Dimensions and Materials

FIGURE 5–9
A Drawing from a Stress Analysis

Critique of the Sample Drawing and Diagram

Figure 5–9: Figure 5–9 is a pair of complementary drawings from a stress analysis. The left-hand drawing shows the shape and critical dimensions of a mechanical assembly and its applied loads. Section arrows (for Section A-A) tell the reader to look for a section through the plane indicated.

A section through the interior of this assembly is shown in the right-hand drawing, which includes thicknesses and drawing numbers with materials or the catalog number and part name. The different symbols in the section help the reader visualize the assembly of the components. In addition to a title at the bottom of the figure, each of the two drawings is labeled to inform the reader of its perspective.

Figure 5–10: Figure 5–10, the critical path for a construction job, is a schematic diagram showing the interdependency of the various construction functions. Each box clearly represents one activity and related information.

A key instructs the reader concerning the information included in each box. The caption at the bottom of the figure informs the reader of its contents.

KEY CONCEPTS

- Graphics add clarity to the written text without extensive description.
- Graphics should be considered as a basic mode of communication for writing reports

STUDENT ASSIGNMENT

In completing the assignments below, perform the following with information and data provided by your instructor:

1. Prepare a sample table of data for your first laboratory experiment.
2. Draw one graph from experimental data for two specimens. Use the full page for your graph and show scattered data points.

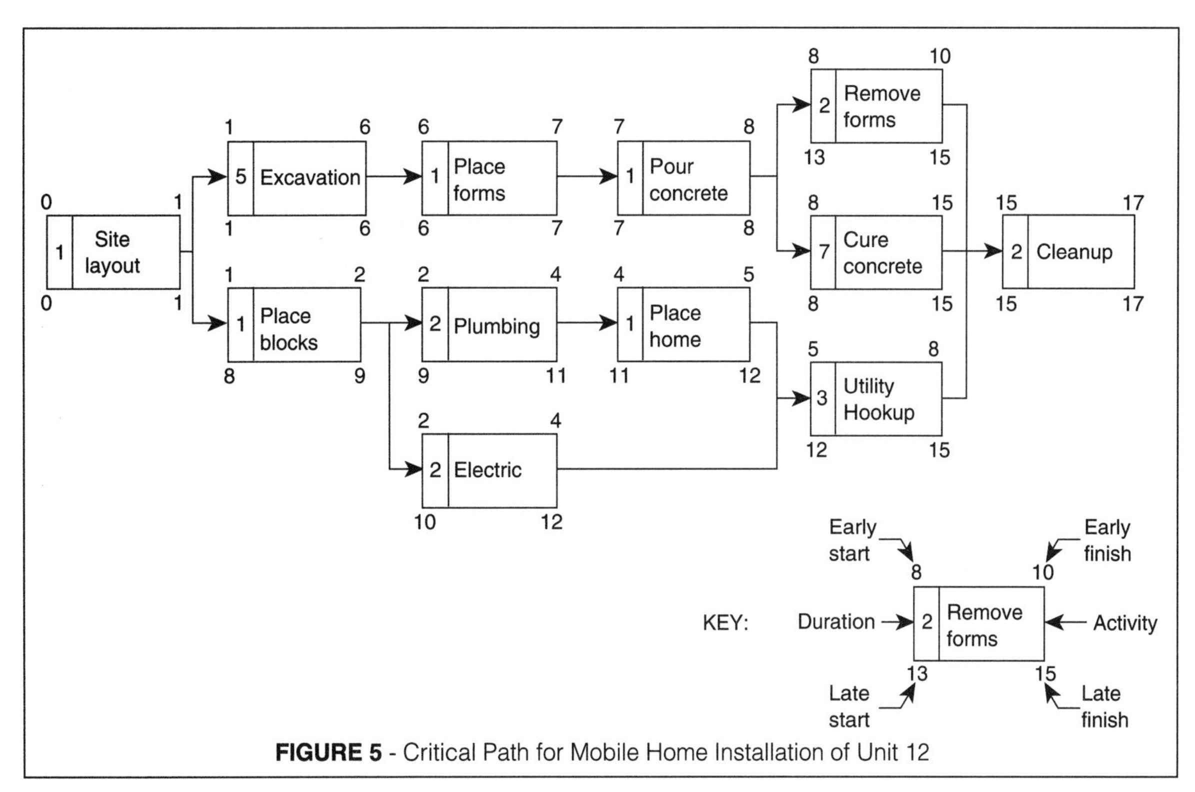

FIGURE 5–10
A Critical Path Diagram

Use approximately one-half of your page for the assignments below:

3. Draw a graph to be used as a designer's technical reference for the same specimens as Assignment 2 above.
4. Prepare a drawing for the equipment and apparatus of a laboratory experiment selected by your instructor. Show envelope dimensions, and major features and controls of importance to an operator.
5. Draw a bar chart for three specimens tested under similar conditions.
6. Draw a pie chart for the results of testing 100 specimens that are in six different categories.

CHAPTER 6

The Title Page and Table of Contents

The title page and table of contents are the envelope of your report and give your instructor a preview of its contents. Instructors receive their first impression from these materials; therefore, these materials should have a professional appearance. The title page and the table of contents are ordinarily prepared after the formal report is completed since they address its contents.

Colleges and universities usually require students to insert their reports in standard laboratory folders to assure that all relevant information is included. These folders ordinarily have a fill-in-the-blank format that is used in lieu of a title page. Also, some instructors do not prefer a table of contents. In this case, the information in this chapter should be superseded by the specific requirements of your instructor.

Most instructors have a preference for binding together the pages and cover of your report. Also, some instructors have preferences for formatting (e.g., margins, line spacing) your report. Check these preferences before submitting your first report to your instructor.

TITLE PAGE

Your instructor sees the title page before the contents of your report; therefore, it should have a neat and crisp appearance. The title page includes the name of your college or university prominently displayed in the upper half of the title page. The department, class number, and section of the laboratory appears directly below the college or university name. Your name may appear either at the top of the page or directly below the class number. The title of the report, ordinarily determined by the instructor, prominently appears directly below the center of the page.

The name of your instructor, the date of the experiment, the date the report is due, and space for a grade may appear in a column in the lower left quadrant of the page. The names of the students, including your own, performing the experiment appears opposite this information in the lower right quadrant of the page. The name of the group leader should be at the top of this otherwise alphabetical list.

The following sample title page includes the above information (Figure 6–1).

Critique of the Sample Title Page

Information relating to the name of the university; the writer of the report; the department, class number and section; the title of the experiment; the instructor's name; the dates of the experiment and the due date; the group number, its members, and leader are clearly visible

MICHAEL JONES

MIDSTATE UNIVERSITY

ME 326-2

MODULUS OF ELASTICITY OF ENGINEERING MATERIALS

INSTRUCTOR:	Prof. N. Chang	Group 3
DATE OF EXPERIMENT:	May 15, 2000	K. Tuan, Leader
DATE DUE:	MAY 22, 2000	L. Greene
		M. Stevens
		C. Wang

GRADE: ___________

FIGURE 6–1
Sample Title Page

to the instructor. A space for the grade is also included. The placement of the information is balanced and esthetically pleasing to the reader.

Your name, the name of the university, class, and title of the experiment are the most important items on the cover page. These are in uppercase letters in the upper half of the page.

Less important items are titled for easy identification and indicated in the lower half of the page.

TABLE OF CONTENTS

A laboratory report longer than four or five pages can include a table of contents to help the reader determine the subject matter of the report, how it is organized, and where to find sections of interest. The order of the sections is usually determined by the instructor.

TABLE OF CONTENTS

List of Tables

List of Figures

FIGURE 6–2
Sample Table of Contents

Major headings and subheadings are labeled using the traditional (1., 1.A., 1.B., 1.C., 2., 2.A., etc.) or the multiple decimal (1.0, 1.1., 1.1.1., 1.2., 2.0, 2.1., etc) systems, and subheadings are indented.

If there are more than four or five tables and figures (sketches, charts, and graphs) in your report, include each of these visual aids in a separate list.

The following sample table of contents is easy to use and is esthetically attractive (Figure 6–2).

In a report, all graphics except tables (e.g., sketches, charts, and graphs are labeled *figures*. Tables are labeled *tables*.

In the table of contents, figures and tables are included in separate lists. The numbering system for each of these visual aids begins at 1, for example, Figure 1, Figure 2, Figure 3, and Table 1, Table 2, Table 3.

The headings and numbering system in the table of contents must correspond identically to the headings and numbering system used in the body of the report.

The sample table of contents in Figure 6–2 is easy to use and is esthetically attractive.

Critique of the Sample Table of Contents

The headings are listed in the order in which the sections are presented and contain the information that identifies the topics of those sections. The list of figures and the list of tables are separate items in the sample table of contents. The appendix is labeled.

Major headings and subheadings are labeled using the multiple decimal system and are indented. (This multiple decimal system and labeled headings in the table of contents must correspond to the multiple decimal system and labeled headings used in the body of the report.) Page numbers are easy to find with a line of dots leading from the headings to the page number. All page numbers are in a right-justified column. Also, the decimal points of heading paragraph numbers and subheading paragraph numbers line up.

The sample table of contents is clearly labeled "Table of Contents" at the top of the page in large, bold letters. The titles for the list of figures and list of tables are in upper- and lowercase letters, respecting the hierarchy of the headings.

KEY CONCEPTS

- The title page and table of contents give previews of the contents of your report.
- Your instructor may judge the quality of the contents of your report based on these materials. Therefore, they must positively impact him or her.

STUDENT ASSIGNMENT

1. Write a cover page and table of contents for the first experiment performed for a laboratory class.

CHAPTER 7

The Beginning of the Report

The beginning of the laboratory report introduces the body of the report. It includes the background information needed to comprehend the material discussed in the report.

The beginning of the report may include the following sections (see Figure 7–1) shown in the order in which they typically appear:

- Summary
- Objective(s), Purpose(s), or Introduction
- Nomenclature and Definitions
- Symbols

Although some instructors require the summary to be at the end of the report, many instructors require it to be placed at the beginning because, except for the conclusions, it is the most important section of the report.

Students usually write the beginning of the report after completing the other parts because it is ordinarily based upon a knowledge and understanding of the information, facts, and data discussed in the body and the ending of the report.

REPORT SUMMARY

The summary is an abbreviated form of the most important parts of the report. The following items are usually addressed in the summary: purpose of the experiment, given facts and data, assumptions, measured data, and results or conclusions. Details are not included.

The summary is probably the most carefully read section of the report and therefore should be carefully written to prevent your instructor from misinterpreting its content. Technical jargon, information, and data are usually kept to a minimum so that your instructor can evaluate your understanding of what was accomplished.

The summary represents the contents of the report and should leave your instructor with the same impression as when the report was read in its entirety. It should only include information that is presented in the body of the report.

The information in the report is briefly reported in the following sample summary. It includes a conclusion and recommendation.

Summary

A torsion test was performed on steel and aluminum specimens to determine their torsional modulii of elasticity. Both members were 2-inch diameter solid shafts that were 12 inches long.

FIGURE 7–1
Sections of the Beginning of the Report

BEGINNING OF THE REPORT
Introduces the body of the report

SUMMARY
A concise statement of the report

OBJECTIVE, PURPOSE, OR INTRODUCTION
Discusses the purpose of the experiment

NOMENCLATURE/DEFINITIONS
A list of terms with special meanings

SYMBOLS
A list of symbols used in the analysis and test

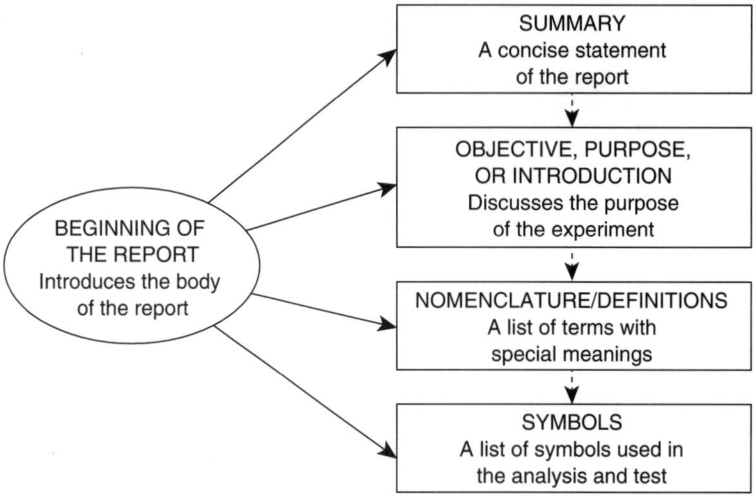

The material properties were assumed to be homogeneous in all directions and that Hooke's law was valid.

The torsional modulii of elasticity was determined to be 12.7×10^6 psi for the steel specimen and 4.5×10^6 psi for the aluminum specimen. These values are 19% and 17% respectively higher than the values published in Brady & Clauser, *Materials Handbook*, 1988, McGraw-Hill.

Critique of the Sample Summary

The summary begins with a statement of purpose. The given facts, data, and assumptions are then indicated. The summary ends with the numerical results of the experiment, and indicates the significance of these results by comparing them to published values.

The methodology, which is not important to the intent or significance, is not included. Notice that, except for details, all pertinent information is included in the summary.

OBJECTIVE(S), PURPOSE(S), OR INTRODUCTION

The objective(s), sometimes called purpose(s), of a laboratory report discusses what you hoped to accomplish with the experiment. When more than one objective is included, they are frequently presented in a list.

The introduction to a laboratory report includes the same information as the objective, but it also discusses how the results of the experiment is hoped to be used in further applications.

The following sample objectives includes what the writer hopes to accomplish with this experiment.

Objectives

The following are the objectives of this flexure test laboratory:

2.1 To determine the flexural modulus of elasticity of a steel beam.
2.2 To determine Poisson's ratio for a steel beam.
2.3 To compare the experimental values of the flexural modulus of elasticity and Poisson's ratio of a steel beam with the theoretical values.

An introduction would add the following concerning how the results of this experiment can be used in further applications:

These data are important in the design of buildings to prevent plaster from cracking by the excessive deflections of ceiling joists.

Critique of the Sample Objectives

An introductory statement identifies the items clearly presented in a list. Each item in the list is clearly independent of the other items and is numbered.

Critique of the Sample Introduction

The introduction adds a statement to indicate how this information is to be used.

NOMENCLATURE AND DEFINITIONS

The nomenclature and definitions section includes a list of terms that have special meanings or are used repeatedly in the report.

Definitions

1. Intermittent weld—a 3-inch fillet weld, 12 inches on center (3–12) unless otherwise specified
2. Pipe—2–inch, Schedule 40 standard steel pipe per ASTM A36
3. Drill—bore a hole with a 1/3-HP drill rotating at 3600 rpm

Critique of the Sample Definitions

These definitions help to simplify the writing when an item or procedure is used repeatedly. For example, whenever the term "pipe" is used in the report, it means "2–inch, Schedule 40 standard steel pipe per ASTM A36."

SYMBOLS

In a laboratory report that has an analytical component, a list of symbols is usually included unless the symbols are standard ones that are commonly used and understood by readers. Symbols are the nomenclature of physical and mathematical quantities and concepts discussed in a report.

Symbols

F_p = proportional elastic limit
F_Y = yield point
F_t = ultimate tensile strength
F_c = ultimate compression strength
μ_e = Poisson's ratio

Critique of the Sample List of Symbols

This list uses standard symbols and is therefore intended for readers not familiar with strength of materials. In the event that the intended readers are familiar with strength of materials, this section of the report can be deleted.

Although it is not recommended, writers sometimes define their own symbols for physical and mathematical quantities and concepts ordinarily represented by standard symbols.

KEY CONCEPTS

- The beginning of the report organizes the material for the reader and facilitates understanding of the body and end of the report.
- The beginning sections are usually written after the body and end of the report are complete.

STUDENT ASSIGNMENT

1. Critique and rewrite the following summary for a tension test report written by a student. Review grammar, syntax, logic, structure, and content.

Research was conducted at the library to make sure that the basic theory and use of the equipment was understood. A test was performed on a column and the load and deformation was recorded.

2. Critique and rewrite the following objective of a deflection test report on wood beams. Review grammar, syntax, logic, structure, and content.

By using the appropriate instrumentation required for the experiment, the test procedure was followed which allowed the student to obtain the fundamental mechanical properties of given structural members. In this case, the characteristics of wood specimens were relevant. Loads are applied by the instrumentation to put three specimens in a bending test, one in a compression test, and one in a shear test. This produces a relation between the loads and the deflection of the specimen by obtaining the appropriate graphical data.

3. Critique and rewrite the following introduction for a strain gage report written by a student. Review grammar, syntax, logic, structure, and content.

The main purpose of this lab experiment was to load a simple steel beam at its non-supported end with a one pound weight in increments of one. To show the strain due to the load, 2 electronic strain gages were attached to the beam and therefore showing the strain of the beam. The strain gage method is very useful in that it can be used to determine the strain of a beam wherein most cases would be impossible to calculate otherwise.

CHAPTER 8

The Body of the Report

The body of the report validates the conclusions. Ordinarily, your instructor will carefully review the body of the report for errors in theory, assumptions, sample calculations, and data and calculations. Therefore, the body of the report must be logically, fully, and accurately presented so that it can be easily understood by your instructor.

The body of the report may include the following sections (see Figure 8–1), typically presented in the order shown below:

- Theory
- Assumptions
- Sample calculations
- List or description of the tested items
- List or description of the equipment and apparatus
- Procedure or process description
- Data and calculations

The body of the report is usually written first as its outcome determines the text of the other parts of the report.

THEORY

A discussion of the theory, including details, should be presented when it is needed to understand the analysis or calculations presented in the report.

The theory can include the derivation of equations, explanations of physical behavior, or both. The entire theory is included in the body of the report only if it can be easily presented. Otherwise, the theory may be included in the appendix (refer to page 72), or by referring the reader to the bibliography.

The theory discussed in the following sample theory is explained in easy-to-understand language.

Theory

For a slender test specimen, small metallurgical grain size of a metal results in low ductility and high strength because slippage of grains is restricted by the presence and resistance of adjacent grains interfering with the slip planes.

FIGURE 8–1
Sections of the Body of the Report

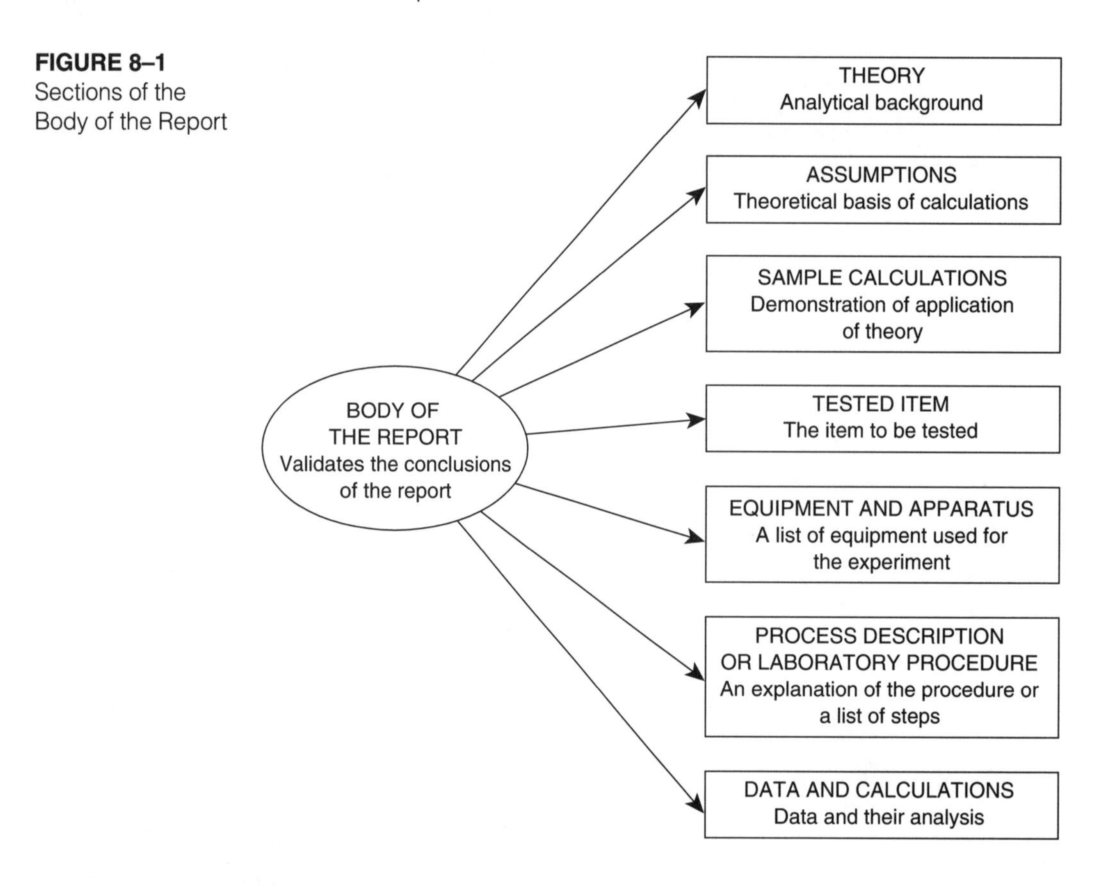

Large metallurgical grain size of a metal results in high ductility and low strength because the centers of adjacent grains are spaced farther apart and the resistance of grains along a plane at an angle to the direction of tension is reduced.

Critique of the Sample Theory

The discussion of physical behavior explains the theory for diametrically opposed conditions (i.e., "small metallurgical grain size" and "Large metallurgical grain size"). Each condition is effectively explained using cause and effect (demonstrated by the use of the phrases *results in* and *because*).

ASSUMPTIONS

If there is an analytical component to the report, it is always based upon certain assumptions. These assumptions are critical to the outcome of the analysis and are clearly specified in a list.

When sample calculations are included with the theory, the assumptions are frequently stated before introducing the sample calculations.

The following sample assumptions, to be used later in the report, are formatted in a list. Each assumption is identified separately.

Assumptions

The following are the assumptions for the calculation of member maximum stresses from the measured truss loads:

1. The load is uniformly distributed across the member cross-section.
2. The modulus of elasticity is 29.2×10^6 psi.
3. The modulus of elasticity of the member is constant along its length.
4. The cross-sectional area of the member is constant along its length.

Critique of the Sample Assumptions

Each of the statements is a premise to be used later in the report. However, the word *is*, the present tense of *to be*, rather than *will*, the future tense, is used. See the section on tense in Chapter 4.

SAMPLE CALCULATIONS

A set of sample calculations demonstrates to your instructor the application of theory and methodology used to analyze the data measured in the experiment. The data and calculations section may be presented as a table or spreadsheet without further explanation. This section is usually included when a set of data are collected and the calculations are repetitive.

A complete discussion for all calculated quantities in the table or spreadsheet that follows the sample calculations must be shown. The assumptions are again introduced where they are required to understand the analysis. Other explanations, such as for the mathematical or physical analysis, are included to simplify the understanding. Usually, students use the first set of collected data to demonstrate the sample calculations.

See pages 31 to 32 for presentation of equations and calculations in laboratory reports.

TECHNICAL DESCRIPTIONS (GENERAL)

Technical descriptions may be included in laboratory reports to describe the tested item and process.

Technical descriptions are used to help your instructor do one of the following:

- Visualize the tested item(s) and, if there are functional components, understand how they function.
- Understand a process.

Use details to help your instructor mentally reconstruct the item, function, and process. Include rates for functions and processes. Only include details that are necessary to reconstruct the item or process. Avoid interesting but unnecessary details. Use precise numbers rather than vague words such as *large*, *fast*, and *very*, which your instructor must interpret.

Use transitional words and phrases to provide continuity for the reader. Use spatial transitional words and phrases such as *above*, *below*, *to the left*, and *adjacent to* for physical descriptions. Use chronological transitional words such as *first*, *second*, *next*, *last*, and *after* for process descriptions.

Analogies explain spatial concepts of size and configuration (e.g., "in the shape of a football 8 inches long and 4 inches in diameter"). Words such as *because*, *therefore*, and *consequently* emphasize cause and effect for function and process descriptions.

Graphics with captions help the reader visualize an item. References to these graphics in your report emphasize the details of the description. Refer to Chapter 5 for more information on using graphics in reports.

Technical Description of an Item

In a technical description of an item, include the following in the order listed:

1. Purpose: a statement of the intended use or function of the item.
2. Physical description: the size, shape, location, and orientation of components, material, color, weight, and other pertinent physical characteristics, and pertinent mechanical properties (e.g., modulus of elasticity). Begin by describing the object as a whole and then develop the details of its components.

For an item with functional components, include the following:

3. Performance characteristics: a performance description (what it does) of the item (e.g., speed, distance, watts, rate of production, and minimum and maximum limits).
4. Functional operation: a functional description (how it works) of the operation and the components necessary to the operation (e.g., gears, cranks, microchips, and chemicals).

Because the performance characteristics depend upon the functional operation, it is often convenient to discuss them together.

Technical Description of a Process

Include the following in the technical description of a process in the order below:

1. Source or purpose: a statement of the cause (e.g., a natural process) or intended use or function of the process (e.g., a man-made process).
2. Physical description: a physical description of the elements that cause the process or the equipment required for the process.
3. Functional operation: a functional description (how it occurs or works) of the process. Include the elements (e.g., water, heat) for a natural process or the components (e.g., gears, cranks, microchips, chemicals) for a man-made process.

The following sample technical description of this natural process (which is easily simulated in a laboratory) emphasizes the concept of the process rather than its physical characteristics.

Description of the Process

Galvanic corrosion occurs when electrical contact occurs between two electrodes—an anode (the electrode supplying the electrons from an external circuit) and a cathode (the electrode receiving the electrons from an external circuit)—in a cell filled with water.

Electrons are released from the anode and flow to the cathode due to the greater electric potential at the anode. The excess electrons at the cathode upset the chemical equilibrium of the system, which liberates hydrogen at the cathode from the hydrogen ions in the water.

This chemical imbalance causes additional electrons to be removed from the anode, which causes a further imbalance of the system. This spontaneous reaction corrodes the anode metal and continues to produce and release hydrogen at the cathode. The excess hydrogen at the cathode plates out in the form of *rust*.

Critique of the Sample Technical Description of a Process

The description begins with discussion of the elements and the environment necessary for the process. The second paragraph discusses the beginning of the process. The description ends with a discussion of how the process continues and the end result.

Similar to the technical description of an item, cause and effect are emphasized.

Chronological transition words are also emphasized, but spatial transition words are conspicuously absent in this process description. The purpose of this description is to help the reader understand the process, not recreate the experiment. Therefore, it is not necessary to describe the physical characteristics since they do not contribute to this understanding.

Words that the reader may not be familiar with are defined in parentheses. This avoids any misunderstanding of terms used in the description. Including the nontechnical word *rust* relates the effect of this process to the reader.

DESCRIPTION OF THE TESTED ITEM

Your instructor will typically require only a list of the items to be tested. However, your instructor may also require a brief physical description of the tested specimen, equipment, structure, or system to confirm its identity.

Soil Samples

Soil samples were obtained from Spadra 323 borrow pit. The unit weight is 110 lb per cu ft in situ.

EQUIPMENT AND APPARATUS

When the laboratory equipment is commonly used or consists of simple components such as strain gages and thermometers, your instructor will typically require only a list of equipment used in the experiment. However, when special equipment is used, your instructor may require you to describe the equipment and apparatus to help you understand its functional characteristics.

Equipment

Rockwell Hardness Tester

The Rockwell hardness tester is a manually operated test machine used to determine the surface hardness of metal specimens. The load is applied to the specimen by an indenter through a system of levers and a counterweight. A dashpot incorporated in the loading system insures gradual loading. A dial gauge is calibrated to read a hardness number, which is inversely proportional to the depth of indentation, i.e., a greater indentation will give a lower hardness number. B, C, and F scales on the dial gauge are each used for materials of different ranges of hardness.

PROCESS DESCRIPTION OR LABORATORY PROCEDURE

Process descriptions are the technical descriptions of laboratory procedures and may be stated in the present or past tense. Alternatively, laboratory procedures use the imperative and include a step-by-step format with step numbers.

The choice of including either a process description or a laboratory procedure depends upon the preference of your instructor.

Process Description

The process description is either an explanation or history of the procedure. Use the present tense for an explanation and the past tense for a history. The following sample process description is an explanation; a history is shown in square brackets ([]). This same event is included later in this section as a procedure for comparison.

Description of Process

The process described below is [was] used to find the cosecant of an angle using an HP-11C calculator.

After turning the calculator on and putting it in the degree mode, the angle is [was] entered into the display from the keyboard. The sine (the reciprocal of cosecant) of the angle is [was] then found. The reciprocal of the value in the display is [was] then determined and read from the display.

Critique of the Sample Process Description

The opening statement tells the reader the purpose of the process. The second paragraph discusses the method used to obtain the desired result.

The second paragraph begins with a chronological transition, *After.* Chronological transitions occur twice more in this paragraph with the word *then.*

Because the reader may not understand the purpose of finding the sine of the angle, the writer explains it. Notice that the explanation of sine is explained with respect to the cosecant (the reciprocal), rather than the functional reason for performing that operation (a cosecant button is not available on the calculator).

Laboratory Procedure

In a laboratory procedure, the sequence of events determines the sequence of the statements.

Example

Instead of writing: Press the STOP button when the load drops abruptly. This may prompt the operator to press the STOP button before there is an abrupt drop in the load.
Write: When the load drops abruptly, press the STOP button.

A sample procedure follows.

How to Find the Cosecant of an Angle on an HP-11C Calculator

1. Press the <**ON**> button.
2. Press <**g**> and <**7**> to put the calculator in the degree mode.
3. Press the digits that represent the angle. Press <**ENTER**>.
4. Press <**SIN**>.
5. Press <**1/x**>.
6. Read the cosecant in the display.

Critique of the Sample Procedure

Each instruction begins with an imperative verb. Step 2 has an explanation to help the reader understand a potentially confusing instruction; otherwise explanations are not given. The names of the buttons are clearly identified in angle brackets. The last step tells the reader to read the cosecant, whose value is the objective of the procedure, in the display.

DATA AND CALCULATIONS

The data are the numerical quantities determined from reading the instruments during the experiment. These are usually recorded on data sheets—tables with the nomenclature of the controlled and measured parameters of the experiment, respectively, as the row and column headings.

The calculations are the analysis of these data and are the numerical quantities used for understanding the results of the experiment. Because of the interrelationship between the experimental data and calculations, the experimental data and calculations are usually shown in the same tables. Frequently, spreadsheets are used for this purpose.

Data and calculations are usually presented in standard formats prepared by your laboratory instructor or included in your laboratory manual. However, when you are required to prepare your own format for tabular information, the data and calculations must be easy-to-follow and include footnoted explanations to facilitate review by your instructor. A bold or double line clearly separates the test data from the calculations.

When the parameters of the experiment are dependent, graphs illustrate the relationship of this dependence. The guidelines on graphics included in Chapter 5 discuss the presentation of tables and graphs. See pages 41 through 46 for a sample data and calculations and its critique.

Some instructors recommend that you place the original data sheets either before the calculations or in the appendix of the report, and that you copy the experimental data into tables that include the calculations.

KEY CONCEPTS

- Technical descriptions should give the reader a clear picture of the subject of the report.
- The analysis is the critical link between the information and data provided and the conclusions and recommendations.

- The ending of the report should be predictable from the information and data presented in the body of the report.

STUDENT ASSIGNMENT

1. Critique and rewrite the following description of the test item. Review grammar, syntax, logic, structure, and content.

> The A36, 16 in. × 4 in. × 1/4 in. steel plate specimens will be used for testing. One specimen will be preheated to 150°F. One specimen will be at room temperature. One specimen will be cooled at 40°F.

2. Critique and rewrite the following equipment and apparatus section. Review grammar, syntax, logic, structure, and content.

> The torsion machine is of the mechanical gear type. It can be operated manually or by an electric motor. The load is transmitted from the moving head to the stationary head, which, in turn, transmits it to the pendulum. The swing of the pendulum balances the applied torque and, at the same time, actuates a scaled lever calibrated to read the torque applied.

3. Critique and rewrite the following process description as an explanation. Review grammar, syntax, logic, structure, and content.

> The centrifugal pump operates at four different constant speeds: approximately 1200, 1400, 1600, and 1800 rpm. At each of these speeds the flow will be varied by means of the discharge valve. Five rates of discharge, in equal increments varying from zero to the maximum, are tested at each of these speeds.
>
> For each discharge rate, the data is measured to compute total head, input horsepower, and output horsepower.

4. Critique and rewrite the following procedure. Review grammar, syntax, logic, structure, and content.

1. Fit the specimen to be tested into spherical end bearings, then measure the unsupported length of the column for center to center of the bearing balls. Measure diameter of the specimen.
2. Balance the scale beam with the procedure indicated on page 14 of its operation manual.
3. Set up the specimen in the testing machine vertically. Adjust the screw sockets to ensure free rotation of the ball bearings. Apply load using the lowest speed of .05 inches per minute.
4. While the load is being applied, the balance lever must be continuously moved outward so as to balance the load. Catch the reading when there is a sudden drop of the scale beam and a simultaneous bowing of the specimen.

5. Critique and rewrite a description of the test item for one of the following as assigned by your instructor:

a. A structural aluminum member
b. An automobile engine or transmission
c. An electronic circuit board
d. A conveyor belt
e. A backhoe, front-loader, or crane
f. A wing of a small airplane
g. Seawater
h. An electric drill

6. Write an equipment and apparatus section for one of the following as assigned by your instructor:
   a. A manometer
   b. A thermometer
   c. A micrometer
   d. A Wheatstone bridge
   e. A balance
   f. A Galvanometer
   g. A gate, globe, or ball valve

7. Write a process description as an explanation or a history for one of the following as assigned by your instructor:
   a. Rotating a television aerial on your roof for best reception
   b. Inflating the air pressure in your tires to 30 psi
   c. Adjusting the right-hand side mirror of your car without remote control
   d. Regulating tap water for comfort
   e. Adjusting lead in a mechanical pencil for drawing
   f. Hanging a picture

8. Write a procedure for one of the processes in Assignment 7, as assigned by your instructor.

CHAPTER 9

The Ending of the Report

A determination of what has been discovered is discussed in the ending of the report. For this reason, it is the most important part of the reporting process. Students inappropriately believing their work is completed when the body of the report is complete and accurate, frequently overlook this importance and write weak endings or omit them altogether. Special emphasis must be placed on writing strong, effective report endings.

When writing conclusions, students need to shift from deductive thinking (reasoning in which a result follows from the stated premises) used in the body of the report to inductive thinking (deriving general principles from a set of facts). The ending of the report is frequently the most difficult to write because students must rely on previous experience and judgment rather than the ability to reason. Therefore, students should discuss the body of the report with their group members before writing the ending.

The ending of the report may include the following sections (see Figure 9–1) shown in the order in which they typically appear:

- Results and conclusions
- Discussion of results
- Bibliography

The appendix is also discussed in this chapter because it follows the report.

RESULTS AND CONCLUSIONS

The results are a summation of the findings of an analysis or a test based on the reasoning presented in the body of the report. The conclusions are the inferences drawn from these findings. The results and conclusions are usually included in the same section (as demonstrated in this chapter) because the results from an analysis or test are frequently what the writer hoped to conclude.

When more than one result and conclusion are presented from the evidence, they should be presented in a list, with results preceding conclusions. Because of its importance, this section is sometimes placed near the beginning of a report.

The organization of the following sample results and conclusions easily translates the results of this laboratory test to the immediate conclusion.

FIGURE 9–1
Sections of the Ending of the Report

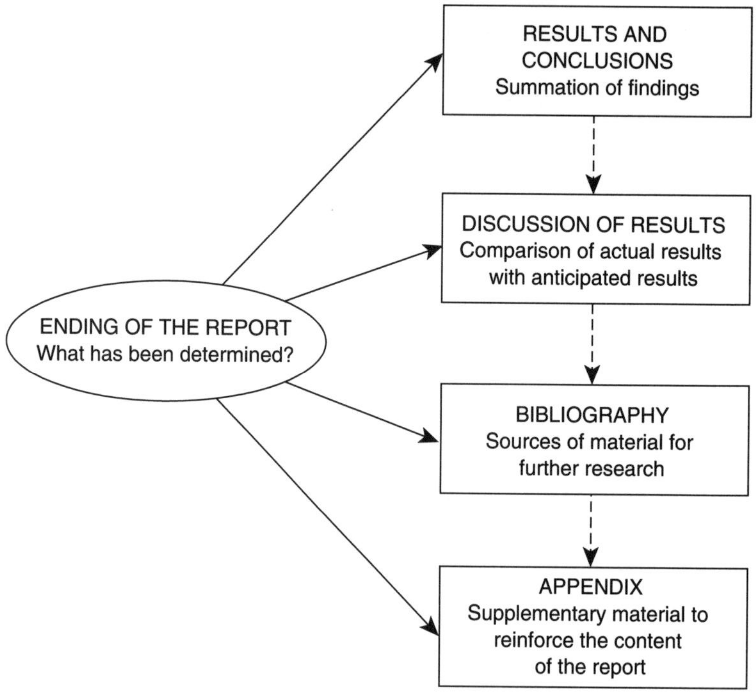

ENDING OF THE REPORT
What has been determined?

RESULTS AND CONCLUSIONS
Summation of findings

DISCUSSION OF RESULTS
Comparison of actual results with anticipated results

BIBLIOGRAPHY
Sources of material for further research

APPENDIX
Supplementary material to reinforce the content of the report

Results and Conclusions

The following are the results from the data and calculations of the efficiency tests:

1. The maximum efficiency is 62% at 450 rpm (see Fig. 4).
2. The specific speed at maximum efficiency is 2.32 (see p. 12).
3. The maximum brake horsepower is 0.60 at 450 rpm (see Fig. 5).

It is therefore concluded that the Model C-80 turbine will convert sufficient kinetic energy at 450 rpm to generate the 0.25 horsepower required to power the Model 361L motor.

Critique of the Sample Results and Conclusions

The results are shown in a list. However, the single conclusion is presented separately as a statement. This statement format facilitates recognizing the difference between the deductive and the inductive reasoning of the writer, while demonstrating the close interrelationship of the conclusion to the results.

Each result includes a reference to the body of the report, demonstrating the relationship of the result to the information in the body of the report.

When there are several conclusions, separate them, and preface the list with a statement similar to "The following are the conclusions of this test determined from the preceding results."

Frequently, a result from the analysis or test is the conclusion. For example, when the purpose of the test in this sample was to determine (1) the maximum efficiency at 450 rpm, (2) the specific speed at maximum efficiency, and (3) the maximum brake horsepower at 450 rpm, these results from the test are also the conclusions, without requiring inductive reasoning to reach them. Therefore, it is common to combine the results and conclusions in the same section.

DISCUSSION OF RESULTS

This section compares the anticipated results with the experimental results obtained in the laboratory. It is indicated when the anticipated results agree with the experimental results. It is also indicated if any revisions in the laboratory procedure were required to obtain the results.

When a laboratory experiment is used to predict behavior, and the behavior is other than predicted, this section may be used to explore the applicability of the experimental procedure.

In the sample discussion of results, the student did not anticipate the problem that was encountered during the experiment. The instructor is presented with sufficient details to understand the explanation.

Discussion of Results

The experimental curves of torque (in.-lb) vs. angle of twist (ϕ) for the cast iron and aluminum specimens do not demonstrate well-defined proportional limits or yield points (see Figure 2). Therefore, the experimental torsional modulii of rigidity, G (psi), cannot be accurately determined. This mechanical behavior is typical of brittle materials. Therefore, to determine the experimental modulii of rigidity, a 0.02% offset is drawn to determine the yield point of each specimen. The torsional modulus of elasticity is then assumed to be the slope of the line extended from the origin to the yield point.

The experimental values of the torsional modulii of elasticity are 19% and 17% higher than the published values, respectively, for the cast iron and aluminum specimens (Ref. 6, p. 83). These values are significantly greater than the ordinary experimental deviation. However, as the extreme fibers of each specimen reach the elastic limit, a combination of increasing plastic deformation of the outer fibers and elastic deformation of the inner fibers react to the applied torsion (see Figure 3). When plastic deformation of the outer fibers begins, the polar moment of inertia, J (in.^{4}), increases as outer fibers become more effective in resisting torsion. This additional polar moment of inertia from the plastic deformation increases the torsional stiffness of the specimen and the apparent *elastic* torsional modulus of rigidity.

To prevent this phenomenon, the torque should be limited to values that are significantly below the proportional limit.

Critique of the Sample Discussion of Results

The first paragraph discusses the revised procedure used to determine the experimental values after a problem is encountered. It states that this problem is typical for the type of material tested.

The second paragraph states that the experimental values are higher than anticipated. Then, the student explains in detail the mechanical behavior (rather than the revised procedure) that is believed to be responsible for the higher values.

The third paragraph discusses a procedure to prevent this undesirable mechanical behavior in the future.

Although the student does not believe that the revised procedure is responsible for the higher values, the revised procedure is included in the discussion. If the student's hypothesis concerning the mechanical behavior is proven incorrect, this revised procedure can then be reviewed for its applicability.

BIBLIOGRAPHY

The bibliography is a list of sources of material for further research by the reader. The information included in the bibliography is supplemental to the information in the report, and should not be relied upon for the reader to fully understand the report.

The citations in a bibliography are alphanumeric and may include page numbers after the name of the publisher.

Bibliography

Davis, Troxel and Hauck, *Testing of Engineering Materials*, 4th Ed., 1992, McGraw-Hill, 68–71.

APPENDIX

An appendix includes supplementary material used to reinforce the material in the report, for example, original data sheets or extensive theory.

Laboratory reports frequently have separate alphabetical or numerical appendices, such as Appendix A, Appendix B, for the different supplementary materials contained in it.

- When any appendix is longer than two or three pages, a title page introduces the appendices section, and a separate title page precedes each appendix with the title Appendix A (or B) centered on the page.

Otherwise, head each appendix on the top of its first page with the title Appendix A (or B).

Page numbers in an appendix are preceded by the letter designation of the appendix. For example, designate the third page in Appendix C as C3.

KEY CONCEPTS

- The results and conclusions should be based on the information and data presented in the body of the report.
- The discussion of results is the primary source of information when behavior or capability are other than predicted.
- Supplementary material in the appendix may be used to reinforce the information in the report.

STUDENT ASSIGNMENT

1. Critique and rewrite the following results and conclusions for a pipe network systems laboratory. Review grammar, syntax, logic, structure, and content.

   The proper design of a piping network is concerned with two design parameters: head-loss and operating pressure. The following can be determined from this experiment:

   At elevations larger than the source elevation, head-losses must be kept at a minimum in order to keep the operating pressure at a comfortable level. At larger elevations, the head-loss contributes to the lowering of the pressure in accordance with the Bernoulli equation. Therefore, larger pipe diameters should be used.

   At elevations smaller than the source elevation, pipes with smaller diameters can be used. Even though the head-loss will increase, the elevation head will counteract the head-loss according to Bernoulli's equation.

   Generally, the ideal piping network should have roughly the same head-loss in each pipe so that big differences in operating pressures between pipe junctions and corners will be avoided.

   In conclusion, piping networks must be analyzed according to the parameters of desired head-loss and operating pressures. Piping systems must be designed in order to achieve an even distribution of head-loss while providing enough operating pressure through all points in the piping system.

2. Critique and rewrite the following discussion of results for the evaluation of a new computer program. Review grammar, syntax, logic, structure, and content.

   The experiment started with calculating the expected load at failure for each specimen, using 7400 psi (bending and compression) and 1000 psi (shear) as the ultimate stress. Four specimens were used in the bending test, two in the shearing test and one in the compression test.

Before any load was applied, a deflectometer kit and flex arm were attached to the Tinius Olsen machine to measure the load being applied to the specimens. In addition, the recorder had to be adjusted and set. Then a load was applied continuously and slowly until the specimen failed. During the entire experiment, the recorder recorded the specimen's process.

From the bending samples, it is interesting to find that the outer areas of the wood, the spaces between the wood fibers are large, which represents the tension caused by the bending load. Small spaces between the wood fibers appear in the inner areas of the wood specimen, caused by compression.

As the duration of the maximum load increased, the strength of the wood decreases (see Figure 1). At one hour, the ratio of working stress to recommended stress for long-time loading is at 165%. It decreases to 130% in one month, then finally to 120% in one year.

CHAPTER 10

A Sample Student Lab Report

The sample student laboratory report demonstrated and critiqued in this chapter shows the structure and content that may be required by your instructor. However, if your instructor requires structure and content other than what is demonstrated in this sample report, you can use the quality of presentation of this sample report as an excellent example.

Figure P–1 (p. vi) shows the *Instructor Requirements for Laboratory Reports* matrix. One column should be completed under the direction of your instructor before you write your first report for that class. A sample laboratory class, ME 299, is included to show how this matrix is filled in. Also, your instructor should inform you of any special requirements not included in this matrix.

GENERAL

The sample laboratory report shows the results of an experiment that includes multiple processes: experimental, analytical, and graphical. Ineffective communication of these dissimilar processes and their results often confuse the instructor and shows that the student did not understand the distinctions between the components of the experiment.

To effectively communicate these processes and their results, this student organizes the report into Parts A and B, each with a different objective; then divides these parts into paragraphs 1., 2., etc.; and then subdivides these paragraphs into subparagraphs a), b), etc. This labeling system enables the instructor to trace a specific process and its result throughout the report by following the labels. Organizing the report in this manner demonstrates to the instructor that the student understood the experiment. More importantly, it helped the student understand the experiment and its results before the report was written.

PRESENTATION

- This report has a professional appearance: it is organized so that information is easily found. It has a title page and headings for the different sections, and its layout is well-designed. There are no spelling or typographical errors.
- The text of the report is written with a word processor. Computer programs are available that have the capability to present data in a spreadsheet, and to fit the best curve to a set of data points. Instructors usually prefer that students use these programs whenever possible.
- The headings are centered and printed in bold for emphasis.

THE COVER MATERIAL

- The information on the cover page is easy to find (see Figure 10–1). The name of the school, department, and course number are shown near the top of the page. The title of the laboratory experiment is printed in bold above the center of the page. The student's name appears below the center of the page. Group members are listed in alphabetical

California State Polytechnic University, Pomona
Civil Engineering Department

CE 342—Hydraulics
LAB 5—VELOCITY AND FLOW MEASUREMENTS

Prepared By
Kimberly Lane

Group 1:	K. Lane, group leader
	C. Jones, T. Nguyen, M. Woo
Date of experiment:	Feb. 11, 2000
Submittal date:	Feb. 18, 2000
Instructor:	D. Wells

FIGURE 10–1
Sample of Cover Page.

order and its leader is identified. All of this information is attractively centered on the page width.

The date of experiment, submitted date, and the name of the instructor are shown in tabular form near the bottom of the page. This student did not include a blank space for a grade. Many instructors prefer a designated location for the student grade.

Most instructors require students to use a standard laboratory report folder instead of a cover page.

- Because the instructor—the only intended reader of this report—has predetermined for the student which sections to include, and will read the entire report, a table of contents is not included. However, for a report more than four or five pages in length, a table of contents adds to the organization.

THE BEGINNING OF THE REPORT

- The Objectives section is to help the instructor understand the purpose of the experiment. Since this experiment has two objectives, this report separates the objectives into Parts A and B (see Figure 10–2).

THE BODY OF THE REPORT

- A list of Apparatus used in the experiment is frequently included in student laboratory reports (see Figure 10–2). However, descriptions are usually not included. Stop watches and scales, as well as items which are commonly found in experimental laboratories, are typically excluded from this list.

OBJECTIVES

PART A: To determine and compare the velocities in an open rectangular channel using a pygmy current meter and a flowmeter. The average velocity is determined from the pygmy current meter measurements.

PART B: To determine and compare the flow rates in an open rectangular channel from data measured using a:

1. rectangular weir
2. v-notch weir
3. normal depth measurement under uniform flow conditions
4. pygmy current meter, and
5. flowmeter

APPARATUS

PART A:

Open rectangular channel
Pygmy current meter
Flowmeter
Scale
Timer

PART B:

Open rectangular channel
Rectangular weir
V-notch weir
Pygmy current meter
Flowmeter
Scale
Timer

1

FIGURE 10–2

- The explanatory Process Description section is presented in Parts A and B for each of the objectives. Each part is divided into paragraphs, each of which describes one component of the experiment (see Figure 10–3). This description discusses the operation that is performed, rather than lists the steps for performing the operation.

PROCESS DESCRIPTION

PART A:

As water flows in an open channel, four different methods are used to determine the average velocity.

1. Using a pygmy current meter, velocities are measured at distances from the water surface of 20%, 40%, 60%, and 80% of the total water depth. The first average velocity is taken as the velocity at the 60% water depth. The second velocity is determined by averaging the measured velocities at the 20% and 80% water depths.

2. The vertical velocity profile is plotted from this data. The third average velocity is graphically determined from this profile.

3. Using a flowmeter to measure the quantity of flow, and knowing the cross-sectional area of the channel, the fourth average velocity is calculated.

PART B:

As water flows in an open channel, quantities of flow are determined by five different methods.

1. Using a rectangular weir first and then a v-notch weir, each upstream depth of flow is measured with a scale. The quantity of flow is calculated using each of these depths.

2. After establishing uniform flow, the normal depth is measured and used to calculate the flow rate using Manning's equation.

3. Using a pygmy current meter in an open channel to measure the velocity at 60% of the depth of flow from the water surface, and knowing the cross-sectional area of the channel, the quantity of flow is calculated.

4. Using a flowmeter located in the pipe leading into the open channel, the quantity of flow is determined.

2

FIGURE 10–3

- Data includes sketches to identify the symbols that are used (see Figure 10–4). The tables include column headings with units. The data is presented in the same sequence that it was collected and each section is clearly identified. Conversion factors are shown in square brackets ([ ]) for identification. An original data sheet may be included in the Appendix.
- The vertical velocity profile graph is clearly titled and includes information concerning the conditions of the experiment (see Figure 10–5). The axes are clearly la-

DATA

PART A – VELOCITY PROFILE:

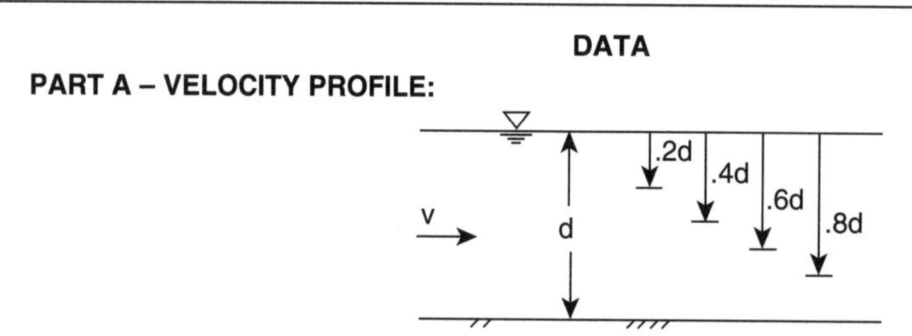

1. PYGMY CURRENT METER

DISTANCE FROM WATER SURFACE	DEPTH FROM BOTTOM OF CHANNEL (in.)	VELOCITY, V (ft./min.)
0.2d	6.0	55
0.4d	4.5	68
0.6d	3.0	66
0.8d	1.5	70

2. FLOWMETER

$$\text{FLOW} = \frac{100 \text{ gal}}{19.21 \text{ sec}} = 5.21 \frac{\text{gal}}{\text{sec}} \left[\frac{60 \text{ sec}}{\text{min}}\right] = 312 \frac{\text{gal}}{\text{min.}}$$

PART B – FLOW:

1.

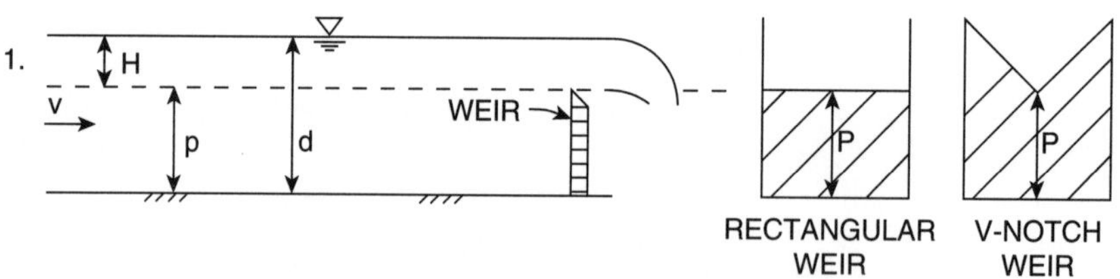

DEPTH	RECTANGULAR WEIR (in.)	V-NOTCH WEIR (in.)
P	6.0	6.0
H	2.4	3.875
d	8.4	9.875

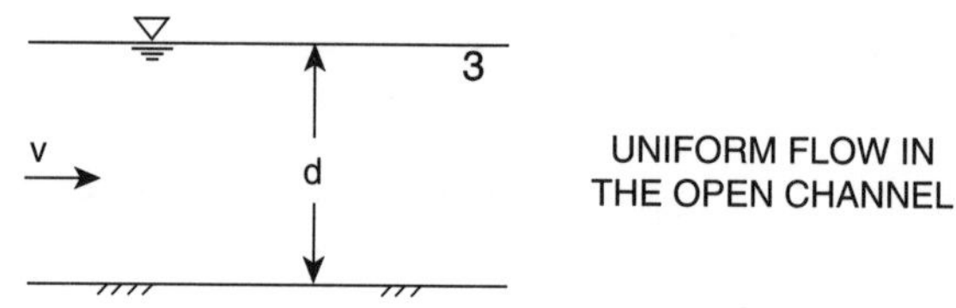

2. PYGMY CURRENT METER IN THE OPEN CHANNEL WITH d = 2.25 in.
   v = 43 ft/min

3. FLOWMETER IN THE INLET PIPE WITH CHANNEL d = 2.25 in.
   Q = 120 gal/min

FIGURE 10–4

beled with units that can be plotted from the data already presented without further calculation or conversion. The divisions of the scales are numbered so that intermediate divisions can be readily interpolated.

- The experimental values of the graph are indicated with dots enclosed by hollow circles. The curve through these experimental values is the best fit of a smooth curve and does not necessarily pass through the center of all the hollow circles. Also, the curve does not cross the outline of any circle.

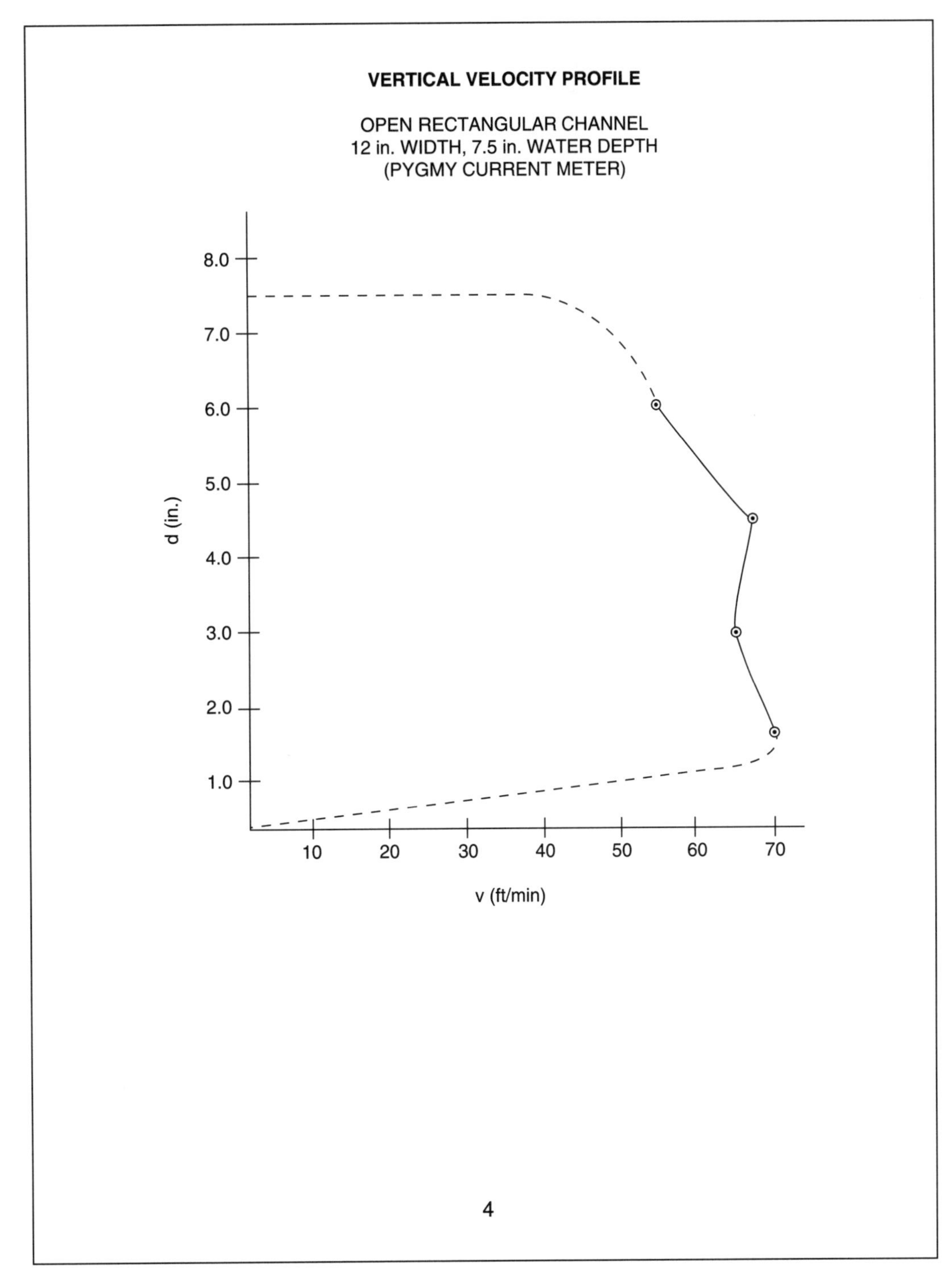

FIGURE 10–5

- The Calculations clearly present the equations used with the data. Each section is labeled for easy identification. Results are underlined and indicated with an arrow drawn from the right margin (see Figure 10–6).
- Explanations are not included in these calculations since the steps are self-explanatory. References are included for uncommon equations.

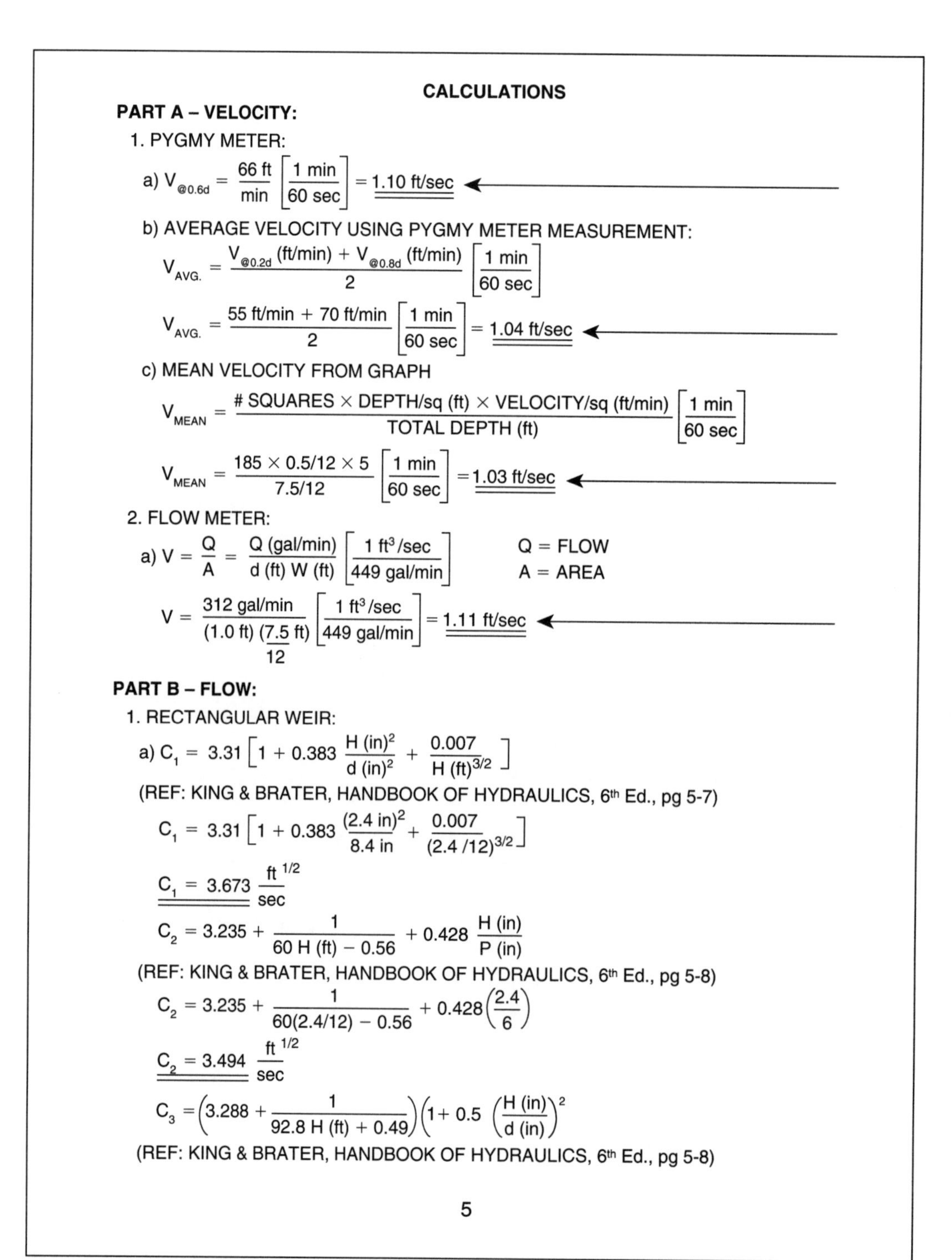

CALCULATIONS

PART A – VELOCITY:

1. PYGMY METER:

a) $V_{@0.6d} = \frac{66\text{ ft}}{\text{min}}\left[\frac{1\text{ min}}{60\text{ sec}}\right] = \underline{\underline{1.10\text{ ft/sec}}}$ ←

b) AVERAGE VELOCITY USING PYGMY METER MEASUREMENT:

$$V_{AVG.} = \frac{V_{@0.2d}\text{ (ft/min)} + V_{@0.8d}\text{ (ft/min)}}{2}\left[\frac{1\text{ min}}{60\text{ sec}}\right]$$

$V_{AVG.} = \frac{55\text{ ft/min} + 70\text{ ft/min}}{2}\left[\frac{1\text{ min}}{60\text{ sec}}\right] = \underline{\underline{1.04\text{ ft/sec}}}$ ←

c) MEAN VELOCITY FROM GRAPH

$$V_{MEAN} = \frac{\text{\# SQUARES} \times \text{DEPTH/sq (ft)} \times \text{VELOCITY/sq (ft/min)}}{\text{TOTAL DEPTH (ft)}}\left[\frac{1\text{ min}}{60\text{ sec}}\right]$$

$V_{MEAN} = \frac{185 \times 0.5/12 \times 5}{7.5/12}\left[\frac{1\text{ min}}{60\text{ sec}}\right] = \underline{\underline{1.03\text{ ft/sec}}}$ ←

2. FLOW METER:

a) $V = \frac{Q}{A} = \frac{Q\text{ (gal/min)}}{d\text{ (ft) } W\text{ (ft)}}\left[\frac{1\text{ ft}^3\text{/sec}}{449\text{ gal/min}}\right]$ Q = FLOW, A = AREA

$V = \frac{312\text{ gal/min}}{(1.0\text{ ft})\left(\frac{7.5}{12}\text{ ft}\right)}\left[\frac{1\text{ ft}^3\text{/sec}}{449\text{ gal/min}}\right] = \underline{\underline{1.11\text{ ft/sec}}}$ ←

PART B – FLOW:

1. RECTANGULAR WEIR:

a) $C_1 = 3.31\left[1 + 0.383\frac{H\text{ (in)}^2}{d\text{ (in)}^2} + \frac{0.007}{H\text{ (ft)}^{3/2}}\right]$

(REF: KING & BRATER, HANDBOOK OF HYDRAULICS, 6th Ed., pg 5-7)

$C_1 = 3.31\left[1 + 0.383\frac{(2.4\text{ in})^2}{8.4\text{ in}} + \frac{0.007}{(2.4/12)^{3/2}}\right]$

$\underline{\underline{C_1 = 3.673\ \frac{\text{ft}^{1/2}}{\text{sec}}}}$

$C_2 = 3.235 + \frac{1}{60\,H\text{ (ft)} - 0.56} + 0.428\frac{H\text{ (in)}}{P\text{ (in)}}$

(REF: KING & BRATER, HANDBOOK OF HYDRAULICS, 6th Ed., pg 5-8)

$C_2 = 3.235 + \frac{1}{60(2.4/12) - 0.56} + 0.428\left(\frac{2.4}{6}\right)$

$\underline{\underline{C_2 = 3.494\ \frac{\text{ft}^{1/2}}{\text{sec}}}}$

$C_3 = \left(3.288 + \frac{1}{92.8\,H\text{ (ft)} + 0.49}\right)\left(1 + 0.5\left(\frac{H\text{ (in)}}{d\text{ (in)}}\right)^2\right)$

(REF: KING & BRATER, HANDBOOK OF HYDRAULICS, 6th Ed., pg 5-8)

5

FIGURE 10–6

CALCULATIONS (CONT'D)

PART B – FLOW: (cont'd)

RECTANGULAR WEIR (cont'd)

$$C_3 = \left(3.288 + \frac{1}{92.8\ H\ (2.4/12) + 0.49}\right)\left(1 + 0.5\ \left(\frac{2.4}{8.4}\right)^2\right)$$

$$\underline{\underline{C_3 = 3.477\ \frac{ft^{1/2}}{sec}}}$$

$$C_{AVG.} = \frac{C_1 + C_2 + C_3}{3}$$

$$C_{AVG.} = \frac{3.673 + 3.494 + 3.477}{3}$$

$$\underline{\underline{C_{AVG.} = 3.548\ \frac{ft^{1/2}}{sec}}}$$

b) WEIR FORMULA:

$$Q = C_{AVG}\left(\frac{ft^{1/2}}{sec}\right) \times\ L\ (ft) = H\ (ft)^{3/2}$$

$$Q = (3.548\ \frac{ft^{1/2}}{sec})(1.0\ ft)\left(\frac{2.4}{12}\ ft\right)^{3/2}$$

$$\underline{\underline{Q = 0.317\ ft^3/sec}} \longleftarrow$$

2. V-NOTCH WEIR

a) WEIR FORMULA:

$$Q = 2.50\ H\ (ft)^{2.48}$$

$$Q = 2.50\left(\frac{3.875}{12}\right)^{2.48}$$

$$\underline{\underline{Q = 0.152\ ft^3/sec}} \longleftarrow$$

3. UNIFORM FLOW IN OPEN CHANNEL

a) $Q = \dfrac{1.486\ S_f^{\ 1/2}\ d\ (ft)^{8/3}}{n}\left[\dfrac{AR^{2/3}}{d\ (ft)^{8/3}}\right]$

where:

S_f = SLOPE OF CHANNEL
n = MANNING'S COEFFICIENT
A = CROSS-SECTIONAL AREA (ft^2)
R = HYDRAULIC RADIUS (ft)
b = WIDTH OF CHANNEL

FOR $\dfrac{d}{b} = \dfrac{2.25}{12} = .188,\ \dfrac{AR^{2/3}}{d^{8/3}} = 4.456$

(REF: JANGER, BASIC HYDRAULICS FOR CIVIL ENGINEERS, 2ND EDITION, JULY 1985, APPENDIX 5-2-1 pg. 573)

$$Q = \frac{(1.486)\ (0.001)^{1/2}}{0.01}\left(\frac{2.25}{2}\right)^{8/3}(4.456)$$

$$\underline{\underline{Q = 0.241\ ft^3/sec}} \longleftarrow$$

6

FIGURE 10–6, *Continued*

CALCULATIONS (CONT'D)

PART B – FLOW: (cont'd)

4. PYGMY METER:

a) $V_{@\ 0.6d} = 89\ \frac{ft}{min} \times \frac{1\ min}{60\ sec} = 1.483\ ft/sec$

b) $Q = V\ (ft/sec) \times\ A\ (ft^2) \times V\ (ft/sec) \times W\ (ft) \times d\ (ft)$

$Q = 1.483\ \frac{ft}{sec}\ (1.0\ ft)\left(\frac{2.25ft}{12}\right)$

$\underline{\underline{Q = 0.278\ ft^3/sec}}$ ←

5. FLOWMETER

a) $Q = \text{FLOWMETER (gal/min)}\left[\frac{1\ ft^3/sec}{449\ gal/min}\right]$

$Q = 120\ \frac{gal}{min}\left[\frac{1\ ft^3/sec}{60\ gal/min}\right]$

$\underline{\underline{Q = 0.267\ ft^3/sec}}$ ←

FIGURE 10–6, *Continued*

THE ENDING OF THE REPORT

- The Summary of Results tabulates the results of the calculations (see Figure 10–7). All results are clearly identified.

SUMMARY OF RESULTS

PART A:

Method of Determination	Velocity (ft/sec)
1. Pygmy current meter at 0.6d	1.10
2. Average velocity calculated using the pygmy current meter at 0.2d and 0.8d	1.04
3. Graph	1.03
4. Flowmeter	1.11

PART B:

Method of Determination	Flow (ft^3/sec)
1. Rectangular weir	0.317
2. V-notch weir	0.152
3. Open channel uniform flow	0.241
4. Pygmy current meter at 0.6d	0.278
5. Flowmeter	0.267

8

FIGURE 10–7

- The Discussion of Results compares the results from the different methods of determination. It also discusses the possible causes of irregularities with the anticipated results (see Figure 10–8).

DISCUSSION OF RESULTS

PART A:
The four different values of flow velocity are determined to be the following:

Method of Determination	*Velocity (fps)*
1. Pygmy current meter at 0.6d	1.10
2. Average velocity calculated using 0.2d and 0.8d	1.04
3. Graph	1.03
4. Flowmeter	1.11

The extreme values among the four velocities differ by 8%.
The vertical velocity profile determined by measurements with the pygmy current meter is an irregular curve (see page 6). Ordinarily, this profile is a smooth curve and demonstrates that the velocity increases nearer to the water surface. This irregularity is probably due to the local turbulence caused by friction at the bottom of the channel and the adverse slope.

PART B:
The five different values of flow quantity are determined to be the following:

Method of Determination	*Flow (cfs)*
1. Rectangular weir	0.317
2. V-notch weir	0.152
3. Open channel uniform flow	0.241
4. Pygmy meter at 0.6d	0.259
5. Flowmeter	0.267

Because the length of the channel is not adequate to establish uniform flow, the depth of flow measured in this experiment is lower than anticipated. Hence, the quantity of flow determined using the uniform flow method, which is dependent upon this measured depth of flow, is less than anticipated, and therefore is less than the flow determined using the rectangular weir and average velocity measurements. The reduction of the measured flow in the v-notch weir was caused by seepage at the sides of the weir.

9

FIGURE 10–8

- The Conclusions discusses the results with respect to the objectives stated at the beginning of the report (see Figure 10–9).

CONCLUSIONS

PART A:

The range of velocities is 8% using the four different methods of determination. This range is within the anticipated accuracy of engineering measurements, revealing that the measured velocity is sufficiently reliable regardless of the method of determination. Although these velocities vary within an acceptable range, the velocity determined by the flowmeter is greater (up to 7%) than the velocities using the pygmy current meter.

PART B:

The open channel flows, theoretically, should be the same for all five methods of determination. The pygmy current meter and flowmeter flows differ by only 4%, which is within the anticipated accuracy of engineering measurements. The flow determined from the rectangular weir measurements was high, whereas, the flow resulting from the v-notch weir was low. This indicates that the weirs are not as accurate as one would expect.

Although the extreme measured flows vary by greater than 50%, the results of the different methods are acceptable approximations. In practice, the simplest method of measurement would typically be used. Because uniform flow was not established, the quantity of flow was not as reliable as when using the other methods of determination.

10

FIGURE 10–9